Mechanical Engineering Essentials with Python: Fluid Mechanics and Hydraulic Systems

Jamie Flux

https://www.linkedin.com/company/golden-dawn-engineering/

Contents

1 Navier-Stokes Equations — 12
    Introduction — 12
    Derivation of Navier-Stokes Equations — 12
    1   Conservation of Mass — 12
    2   Conservation of Momentum — 13
    3   Constitutive Relations and Stress Tensor — 13
    4   Final Formulation — 13
    Applications in Incompressible Flows — 13
    1   Laminar Flow in Pipes — 13
    2   Flow Around Objects — 14
    Numerical Solutions — 14
    1   Finite Volume Method — 14
    2   Computational Fluid Dynamics (CFD) — 14
    Experimental Validation — 14
    Python Code Snippet — 14
    Multiple Choice Questions — 17

2 Reynolds-Averaged Navier-Stokes (RANS) — 20
    Introduction to RANS Equations — 20
    Derivation of RANS Equations — 20
    Reynolds Stress Tensor — 21
    Closure Models — 21
    1   The Boussinesq Hypothesis — 21
    2   `k-epsilon` Model — 21
    Simulation Techniques for RANS — 22
    1   Finite Volume Method (FVM) — 22
    2   Numerical Solvers — 22
    Applications in Engineering — 22
    1   Aerodynamics — 22
    2   Hydraulic Systems — 22

        Python Code Snippet 23
        Multiple Choice Questions 26

3 Turbulence Kinetic Energy Equation   29
        Introduction to Turbulence Kinetic Energy (TKE) . 29
        Derivation of the TKE Equation 29
        Production of Turbulence Kinetic Energy 30
        Dissipation of Turbulence Kinetic Energy 30
        Equilibrium and Imbalance Dynamics 30
        Numerical Implementation of TKE Equation 31
        Applications in Engineering System Analyses 31
        Python Code Snippet 31
        Multiple Choice Questions 33

4 Bernoulli's Equation - Extended Application   36
        Foundation of Bernoulli's Equation 36
        Extended Bernoulli's Equation in Varying Cross-Sections 36
        Unsteady Flow Considerations 37
        Practical Applications in Hydraulic Systems 37
        Mathematical Modelling in Computational Fluid Dynamics 38
        Python Code Snippet 38
        Multiple Choice Questions 40

5 Vorticity Transport Equation   43
        Vorticity in Fluid Mechanics 43
        Derivation of the Vorticity Transport Equation . . . 43
        Vortex Dynamics and Stretching 44
        Implications for Turbulent Flows 44
        Mathematical Modeling 45
        Python Code Snippet 45
        Multiple Choice Questions 48

6 Potential Flow Theory   51
        Governing Equations and Assumptions 51
        Boundary Conditions 51
        Stream Function Formulation 52
        Complex Potential and Conformal Mapping 52
        Applications in External Flows 53
        1   Flow Around a Cylinder 53
        2   Source and Sink Flows 53
        Limitations and Idealizations 53

  Python Code Snippet 54
  Multiple Choice Questions 56

7 Pressure Poisson Equation   59
  Derivation of the Pressure Poisson Equation 59
  Numerical Implementation in Fluid Simulations . . . 60
  Boundary Conditions for Pressure Fields 60
  Role in Navier-Stokes Solvers 60
  Advanced Topics: Coupling with Turbulence Models 61
  Python Code Snippet 61
  Multiple Choice Questions 64

8 Energy Equations for Compressible Fluids   67
  Governing Equations 67
  Internal Energy Models 67
  Heat Transfer Modeling 68
  Work Interactions 68
  Entropy Considerations 68
  Numerical Approaches 69
  Python Code Snippet 69
  Multiple Choice Questions 72

9 The Euler Equations   75
  Introduction to the Euler Equations 75
  Assumptions and Simplifications 75
  Applications in Aerospace Engineering 76
  Relevance to Aerodynamic Design 76
  Numerical Simulation Techniques 77
  Python Code Snippet 77
  Multiple Choice Questions 79

10 Boundary Layer Equations   82
  Introduction to Boundary Layer Theory 82
  Formulation of Laminar Boundary Layer Equations . 82
  Turbulent Boundary Layer Formulation 83
  Similarity Solutions for Laminar Boundary Layers . 83
  Momentum Integral Equation in Boundary Layers . 83
  Analysis and Estimation of Turbulent Boundary Layers 84
  Computational Techniques for Boundary Layer Analysis 84
  Python Code Snippet 85

      Multiple Choice Questions 87

11 Conservation of Vorticity   90
      Introduction to Vorticity Dynamics 90
      Vorticity Equation Derivation 90
      Implications of Vorticity Conservation 91
      Vortex Stretching and Tilting 91
      Role in Turbulent Flows 91
      Applications in Engineering 91
      Python Code Snippet 92
      Multiple Choice Questions 94

12 K- (K-Epsilon) Turbulence Model   97
      Introduction to Turbulence Modeling 97
      Turbulent Kinetic Energy Equation 97
      Dissipation Rate Equation 98
      Closure Coefficients and Turbulent Viscosity 98
      Application in Complex Fluid Systems 98
      Limitations and Opportunities for Improvement . . . 99
      Python Code Snippet 99
      Multiple Choice Questions 101

13 K- (K-Omega) Turbulence Model   104
      The K- Model Formulation 104
      Near-Wall Turbulence Modeling 105
      Empirical Constants and Turbulent Viscosity 105
      Application and Numerical Stability 105
      Python Code Snippet 106
      Multiple Choice Questions 109

14 Momentum Integral Equation   111
      Introduction to Momentum Integral Equation 111
      Displacement and Momentum Thickness 111
      Derivation of the Momentum Integral Equation . . . 112
      Application in Boundary Layer Prediction 112
      1   Utilization in Laminar and Turbulent Flows . 113
      Numerical Methods and Computational Applications 113
      Python Code Snippet 113
      Multiple Choice Questions 115

15 Cavitating Flow Dynamics — 118
- Introduction to Cavitation Phenomena 118
- Governing Equations in Cavitating Flows 118
- Vapor Bubble Dynamics 119
  - 1   Nucleation and Growth 119
- Implications in Hydraulic Systems 119
  - 1   Experimental and Computational Studies . . 120
- Python Code Snippet 120
- Multiple Choice Questions 122

16 Multiphase Flow Quantification — 125
- Governing Equations for Multiphase Flow 125
  - 1   Continuity Equation 125
  - 2   Momentum Equation 125
  - 3   Energy Equation 126
- Modeling Strategies for Multiphase Flows 126
  - 1   Volume of Fluid (VOF) Method 126
  - 2   Eulerian-Eulerian Models 126
  - 3   Lagrangian Approaches 127
- Simulation Techniques 127
  - 1   Finite Volume Method (FVM) 127
  - 2   Coupling Mechanisms 127
  - 3   Turbulence Models 127
- Python Code Snippet 128
- Multiple Choice Questions 131

17 Compressible Flow – Isentropic Relations — 134
- Fundamentals of Compressible Flow 134
- Isentropic Flow Relations 134
  - 1   Temperature-Pressure Relationship 135
  - 2   Density-Temperature Relationship 135
- Application in Engineering Systems 135
  - 1   Nozzle Design 136
  - 2   Diffuser Performance 136
  - 3   Supersonic Flow Analysis 136
- Isentropic Flow Realization and Limitations 136
- Python Code Snippet 137
- Multiple Choice Questions 139

18 Hydraulic Jump Analysis — 142
- Introduction to Hydraulic Jumps in Open Channels — 142
- Fundamental Equations of Hydraulic Jumps — 142
  - 1 Continuity Equation — 142
  - 2 Momentum Equation — 143
- Energy Dissipation in Hydraulic Jumps — 143
  - 1 Energy Loss Computation — 143
  - 2 Relation Between Upstream and Downstream Depths — 143
- Applications and Implications in Engineering Systems — 144
- Python Code Snippet — 144
- Multiple Choice Questions — 146

19 Hydrodynamic Stability Theory — 149
- Introduction to Hydrodynamic Stability — 149
- Governing Equations of Stability Analysis — 149
  - 1 Linearization of the Navier-Stokes Equations — 149
  - 2 Normal Mode Analysis — 150
- Rayleigh's Stability Equation for Inviscid Flows — 150
- Orr-Sommerfeld Equation for Viscous Flows — 150
- Applications in Fluid Interfaces — 151
  - 1 Kelvin-Helmholtz Instability — 151
  - 2 Rayleigh-Taylor Instability — 151
- Complex Variable Theory in Stability Analysis — 151
- Python Code Snippet — 152
- Multiple Choice Questions — 154

20 Laminar Flow: Exact Solutions — 157
- Introduction to Laminar Flow Regimes — 157
- Governing Equations for Laminar Flow — 157
  - 1 Navier-Stokes Equations for Incompressible Flow — 157
  - 2 Continuity Equation for Incompressible Flow — 158
- Exact Solutions in Simple Geometries — 158
  - 1 Hagen-Poiseuille Flow in Circular Pipes — 158
  - 2 Couette Flow Between Parallel Plates — 158
- Laminar Flow Over a Flat Plate: Blasius Solution — 158
- Stokes' First Problem: Plate Sudden Motion — 159
- Laminar Flow Through Channels: Pouseuille and Plane Couette Plates — 159
  - 1 Plane Poiseuille Flow — 159
  - 2 Superposed Couette and Poiseuille Flow — 160

    Complex Geometries: Limitations and Extensions . . 160
    Python Code Snippet 160
    Multiple Choice Questions 162

21 Wave Motion Equations in Fluids   165
    Foundation of Wave Motion in Fluid Dynamics . . . 165
    Governing Equations for Linear Waves 165
    1   Linear Wave Equation 165
    2   Dispersion Relations for Linear Waves 166
    Analysis of Nonlinear Wave Phenomena 166
    1   Korteweg-de Vries (KdV) Equation 166
    2   Nonlinear Schrödinger Equation 166
    Wave Propagation in Stratified Fluids 166
    1   Internal Wave Dynamics 166
    2   Boussinesq Approximation 167
    Numerical Approaches for Complex Wave Systems . 167
    1   Finite Difference Methods 167
    2   Spectral Methods 167
    Applications of Wave Motion Equations in Engineering 167
    Python Code Snippet 168
    Multiple Choice Questions 171

22 Pipe Networks Hydraulic Analysis   174
    Introduction to Hydraulic Networks 174
    Fundamental Equations for Node Analysis 174
    1   Continuity Equation at Nodes 174
    2   Energy Equation in Loops 175
    Head Loss Calculations 175
    1   Darcy-Weisbach Equation 175
    2   Empirical Correlations for Friction Factor . . 175
    Optimization Techniques in Pipeline Systems 176
    1   Flow Distribution Optimization 176
    2   Linear Programming for Network Optimization 176
    Numerical Simulation of Pipe Networks 176
    1   Finite Element Analysis 176
    2   Computational Fluid Dynamics (CFD) Approaches 177
    Python Code Snippet 177
    Multiple Choice Questions 179

23 Stream Function and Complex Potential · 182
Two-Dimensional Flow Analysis 182
    1   Stream Function Dynamics 182
    2   Complex Potential Applications 183
Conformal Mapping Techniques 183
    1   Mathematical Formulation of Conformal Maps 183
    2   Applications in Engineering Problems 184
Practical Implications of Stream Function and Complex Analysis 184
    1   Flow Around Bodies and Obstacles 184
    2   Integration into Computational Methods . . . 184
Python Code Snippet 185
Multiple Choice Questions 187

24 Flow-Induced Vibrations · 190
Introduction to Flow-Induced Vibrations 190
Governing Equations 190
Fluid-Structure Interaction Mechanisms 191
    1   Vortex-Induced Vibrations 191
    2   Flutter Phenomena 191
    3   Buffeting and Unsteady Aerodynamics 192
Impact on Structural Design 192
Python Code Snippet 192
Multiple Choice Questions 194

25 Inviscid Irrotational Flow Theory · 197
Theoretical Foundations of Inviscid Flow 197
Irrotational Flow and Potential Function Theory . . 197
Bernoulli's Equation in Potential Flow 198
Stream Function and its Properties 198
Complex Potential and Conformal Mapping 198
Flow Patterns and the Superposition Principle . . . 199
Applications in Engineering Analysis 199
Python Code Snippet 199
Multiple Choice Questions 201

26 Mach Number and Flow Regimes · 204
Introduction to the Mach Number 204
Speed of Sound and Thermodynamic Considerations 204
Flow Regimes Classified by Mach Number 205
    1   Subsonic Flow Dynamics 205
    2   Transonic Flow Characteristics 205

    3   Supersonic Flow and Shock Waves 205
    4   Hypersonic Flow Observations 206
    Applications and Numerical Methods 206
    Python Code Snippet 206
    Multiple Choice Questions 208

27 Non-Newtonian Fluid Dynamics   211
    Introduction to Non-Newtonian Fluids 211
    Theoretical Background and Governing Equations . 211
    1   Power-Law Model 212
    2   Bingham Plastic Model 212
    Rheological Measurement Techniques 212
    1   Flow Curve Analysis 212
    Computational Fluid Dynamics Application 212
    1   Discretization Methods 213
    2   Turbulence Models for Non-Newtonian Fluids 213
    Practical Engineering Applications 213
    1   Polymer Processing 213
    2   Biological and Medical Applications 214
    Python Code Snippet 214
    Multiple Choice Questions 217

28 Shock Wave and Expansion Fan Dynamics   220
    Introduction to Supersonic Flows 220
    Shock Wave Fundamentals 220
    1   Normal and Oblique Shock Waves 221
    Expansion Fan Dynamics 221
    1   Mach Waves and Characteristics 222
    Mathematical Formulations in Shock-Expansion Theory 222
    1   CFD Implementation 222
    Applications in Supersonic Engineering 223
    Python Code Snippet 223
    Multiple Choice Questions 225

29 Fluid Flow in Porous Media   228
    Governing Equations for Porous Media Flow 228
    Permeability and Porosity Analysis 229
    Application in Civil and Geotechnical Engineering . 229
    Numerical Modeling in Porous Media 230
    Advanced Topics in Reactive and Multiphase Flows . 230
    Python Code Snippet 231

Multiple Choice Questions 234

30 Heat Exchanger Performance Equations — 237
Introduction 237
Governing Equations 237
    1   The Log Mean Temperature Difference (LMTD) 238
    2   Effectiveness-NTU Method 238
Heat Transfer Coefficients 238
Modeling Temperature Profiles 239
    1   Counterflow and Parallel Flow Configurations 239
Advanced Approaches and Computational Methods . 239
    1   Utilization of CFD in Heat Exchanger Design 240
    2   Optimization Techniques 240
Python Code Snippet 240
Multiple Choice Questions 243

31 Viscoelastic Fluid Flow — 246
Introduction to Viscoelasticity 246
Governing Equations of Viscoelastic Fluids 246
    1   Constitutive Models 246
    2   Stress Tensor and Rate of Deformation . . . 247
Behavior under Deformation and Varying Strain Rates 247
    1   Generalized Maxwell Model 247
    2   Power Law Model for Non-Newtonian Behavior 248
Numerical Simulation Techniques 248
    1   Finite Element Analysis 248
    2   CFD and Viscoelasticity 248
Applications in Engineering Systems 249
Python Code Snippet 249
Multiple Choice Questions 251

32 Two-Phase Flow Instability — 255
Introduction to Two-Phase Flow 255
Stability Criteria in Two-Phase Flow 255
Pattern Formation and Flow Regimes 256
    1   Weber Number 256
    2   Froude Number 256
Mathematical Models of Instability 256
    1   Homogeneous Equilibrium Model 257
    2   Drift-Flux Model 257
Numerical Simulation of Two-Phase Flow Instability 257
    1   Volume of Fluid Method 257

2	Level-Set Method	258
	Python Code Snippet	258
	Multiple Choice Questions	261

33 Piping Systems Surge Analysis   264

	Introduction to Surge Phenomena	264
	Governing Equations	264
1	Continuity Equation	264
2	Momentum Equation	265
	Wave Propagation in Pipelines	265
	Methods for Surge Mitigation	265
1	Pressure Relief Valves	265
2	Air Chambers and Surge Tanks	266
	Numerical Approaches	266
1	Method of Characteristics	266
2	Finite Element Method	266
	Advanced Simulation Techniques	267
	Python Code Snippet	267
	Multiple Choice Questions	271

Chapter 1

Navier-Stokes Equations

Introduction

The Navier-Stokes equations are fundamental to the understanding of fluid motion. These equations provide mathematical models that describe the physics of fluid motion, particularly in the fields of fluid dynamics and fluid mechanics. The equations are derived from the fundamental principles of conservation of mass, momentum, and energy in a fluid medium.

Derivation of Navier-Stokes Equations

The derivation of the Navier-Stokes equations begins with the application of Newton's second law to fluid motion, coupled with the assumptions of continuum mechanics.

1 Conservation of Mass

The conservation of mass, or the continuity equation, for an incompressible fluid is expressed as:

$$\nabla \cdot \mathbf{u} = 0$$

where $\mathbf{u}$ is the velocity vector field.

2 Conservation of Momentum

The general form of the conservation of momentum is given by:

$$\rho\left(\frac{\partial \mathbf{u}}{\partial t} + (\mathbf{u} \cdot \nabla)\mathbf{u}\right) = -\nabla p + \nabla \cdot \tau + \mathbf{f}$$

where ρ is the fluid density, p is the pressure, τ represents the stress tensor, and $\mathbf{f}$ is the body force per unit volume acting on the fluid.

3 Constitutive Relations and Stress Tensor

For a Newtonian fluid, the stress tensor τ is related to the strain rate tensor through:

$$\tau = \mu(\nabla \mathbf{u} + (\nabla \mathbf{u})^T) - \frac{2}{3}\mu(\nabla \cdot \mathbf{u})\mathbf{I}$$

where μ is the dynamic viscosity and $\mathbf{I}$ is the identity tensor.

4 Final Formulation

Substituting the stress tensor into the momentum conservation equation, we obtain the Navier-Stokes equation for an incompressible Newtonian fluid:

$$\rho\left(\frac{\partial \mathbf{u}}{\partial t} + (\mathbf{u} \cdot \nabla)\mathbf{u}\right) = -\nabla p + \mu \nabla^2 \mathbf{u} + \mathbf{f}$$

Applications in Incompressible Flows

The Navier-Stokes equations are extensively utilized to model incompressible flow scenarios.

1 Laminar Flow in Pipes

For laminar flow in circular pipes, the Navier-Stokes equations can be simplified to derive the Hagen-Poiseuille equation, which describes the volumetric flow rate Q:

$$Q = \frac{\pi R^4 \Delta p}{8\mu L}$$

where R is the pipe radius, Δp is the pressure drop, μ is the dynamic viscosity, and L is the length of the pipe.

2 Flow Around Objects

The analysis of flow around objects, such as airfoils or submerged bodies, often employs the Navier-Stokes equations to predict drag and lift forces. In these applications, the pressure distribution and shear stress on the surface of the object are of particular interest.

Numerical Solutions

Due to the complex nature of the Navier-Stokes equations, analytical solutions are rare. Thus, numerical methods are frequently employed to solve these equations in practical applications.

1 Finite Volume Method

The finite volume method divides the domain into discrete control volumes and applies the integral form of the conservation equations. This method is advantageous for complex geometries.

2 Computational Fluid Dynamics (CFD)

With advancements in computational technologies, CFD software packages allow for the simulation of complex fluid flows by solving the Navier-Stokes equations numerically.

Experimental Validation

Laboratory experiments are crucial to validate the derived models and numerical simulations of the Navier-Stokes equations. Controlled experimental setups enable the detailed study of fluid phenomena under the Navier-Stokes framework.

Python Code Snippet

Below is a Python code snippet that encapsulates the fundamental computational elements involved in solving the Navier-Stokes equations numerically using the finite volume method for simulating incompressible flow scenarios such as flow in pipes and around objects.

```python
import numpy as np
import matplotlib.pyplot as plt

def initialize_domain(nx, ny, viscosity, density):
    '''
    Initialize the computational domain and relevant parameters.
    :param nx: Number of grid points in x-direction.
    :param ny: Number of grid points in y-direction.
    :param viscosity: Kinematic viscosity of the fluid.
    :param density: Density of the fluid.
    :return: Initialized velocity fields, pressure, and other
    ↪ parameters.
    '''
    u = np.zeros((nx, ny))    # Initial x-velocity
    v = np.zeros((nx, ny))    # Initial y-velocity
    p = np.zeros((nx, ny))    # Pressure field
    dx = dy = 0.01            # Grid spacing
    dt = 0.001                # Time step size for the simulation
    return u, v, p, dx, dy, dt, viscosity, density

def solve_navier_stokes(u, v, p, dx, dy, dt, viscosity, density,
↪ nit=50):
    '''
    Solve the Navier-Stokes equations using a simple finite volume
    ↪ method.
    :param u: X-component of the velocity field.
    :param v: Y-component of the velocity field.
    :param p: Pressure field.
    :param dx: Grid spacing in x-direction.
    :param dy: Grid spacing in y-direction.
    :param dt: Time step size.
    :param viscosity: Kinematic viscosity of the fluid.
    :param density: Density of the fluid.
    :param nit: Number of iterations for pressure Poisson equation.
    '''
    nx, ny = u.shape
    for n in range(100):    # Example time-stepping loop
        un = u.copy()
        vn = v.copy()

        # Pressure Poisson equation
        for it in range(nit):
            pn = p.copy()
            p[1:-1, 1:-1] = ((dy**2 * (pn[1:-1, 2:] + pn[1:-1, :-2])
            ↪ +
                             dx**2 * (pn[2:, 1:-1] + pn[:-2, 1:-1])
            ↪ -
                             density * (dy**2 * dx**2) / (2 *
            ↪ (dx**2 + dy**2)) *
                             ((1/dt) * ((un[1:-1, 2:] - un[1:-1,
            ↪ :-2]) / (2 * dx) +
```

```
                            (vn[2:, 1:-1] - vn[:-2,
                         ↪   1:-1]) / (2 * dy)) -
                        ((un[1:-1, 2:] - un[1:-1, :-2]) / (2
                         ↪   * dx))**2 -
                        2 * ((un[2:, 1:-1] - un[:-2, 1:-1]) /
                         ↪   (2 * dy) *
                            (vn[1:-1, 2:] - vn[1:-1, :-2]) /
                         ↪   (2 * dx)) -
                        ((vn[2:, 1:-1] - vn[:-2, 1:-1]) / (2
                         ↪   * dy))**2)))

        # Velocity field updates
        u[1:-1, 1:-1] = (un[1:-1, 1:-1] -
                        un[1:-1, 1:-1] * dt / dx *
                        (un[1:-1, 1:-1] - un[1:-1, :-2]) -
                        vn[1:-1, 1:-1] * dt / dy *
                        (un[1:-1, 1:-1] - un[:-2, 1:-1]) -
                        dt / (2 * density * dx) * (p[1:-1, 2:] -
                         ↪   p[1:-1, :-2]) +
                        viscosity * (dt / dx**2 *
                        (un[1:-1, 2:] - 2 * un[1:-1, 1:-1] +
                         ↪   un[1:-1, :-2]) +
                        dt / dy**2 *
                        (un[2:, 1:-1] - 2 * un[1:-1, 1:-1] +
                         ↪   un[:-2, 1:-1])))

        v[1:-1, 1:-1] = (vn[1:-1, 1:-1] -
                        un[1:-1, 1:-1] * dt / dx *
                        (vn[1:-1, 1:-1] - vn[1:-1, :-2]) -
                        vn[1:-1, 1:-1] * dt / dy *
                        (vn[1:-1, 1:-1] - vn[:-2, 1:-1]) -
                        dt / (2 * density * dy) * (p[2:, 1:-1] -
                         ↪   p[:-2, 1:-1]) +
                        viscosity * (dt / dx**2 *
                        (vn[1:-1, 2:] - 2 * vn[1:-1, 1:-1] +
                         ↪   vn[1:-1, :-2]) +
                        dt / dy**2 *
                        (vn[2:, 1:-1] - 2 * vn[1:-1, 1:-1] +
                         ↪   vn[:-2, 1:-1])))

def visualize_flow(u, v):
    '''
    Plot the velocity field to visualize the fluid flow.
    :param u: X-component of the velocity field.
    :param v: Y-component of the velocity field.
    '''
    plt.figure(figsize=(11, 7), dpi=100)
    plt.quiver(u[::2, ::2], v[::2, ::2])
    plt.title('Velocity field')
    plt.xlabel('X')
    plt.ylabel('Y')
    plt.show()
```

```
# Example usage
nx, ny = 41, 41
viscosity = 0.1
density = 1.0

u, v, p, dx, dy, dt, viscosity, density = initialize_domain(nx, ny,
↪   viscosity, density)
solve_navier_stokes(u, v, p, dx, dy, dt, viscosity, density)
visualize_flow(u, v)
```

This Python code equips users with a basic implementation for numerically solving the Navier-Stokes equations with the following core functions:

- `initialize_domain`: Prepares the computational domain and initializes the velocity and pressure fields as well as essential parameters like viscosity and density.

- `solve_navier_stokes`: Executes the finite volume method to iteratively solve the Navier-Stokes equations, implementing a simple scheme for pressure and velocity updates.

- `visualize_flow`: Leverages matplotlib to visualize the calculated velocity field and comprehend fluid motion, offering key insights into flow dynamics.

The code, which demonstrates an elementary simulation of incompressible flow, is designed to provide a practical basis for more advanced computational fluid dynamics analyses.

Multiple Choice Questions

1. What fundamental equations form the basis of fluid mechanics?

    (a) Bernoulli's equation

    (b) Navier-Stokes equations

    (c) Poisson's equation

    (d) Euler's equations

2. The Navier-Stokes equations are derived based on which principle?

    (a) Conservation of energy

(b) Conservation of entropy

   (c) Conservation of mass and momentum

   (d) Conservation of angular momentum

3. In the context of the Navier-Stokes equations for incompressible fluids, what does the continuity equation signify?

   (a) $\nabla \times \mathbf{u} = 0$

   (b) $\nabla \cdot \mathbf{u} = 0$

   (c) $\nabla \mathbf{u} = 0$

   (d) $\nabla^2 \cdot \mathbf{u} = 0$

4. Which mathematical term in the Navier-Stokes equations generally represents viscous effects in a Newtonian fluid?

   (a) $\nabla \cdot \mathbf{u}$

   (b) $-\nabla p$

   (c) $\mu \nabla^2 \mathbf{u}$

   (d) $(\mathbf{u} \cdot \nabla)\mathbf{u}$

5. For laminar flow in pipes, derivations from the Navier-Stokes equations lead to which law that relates volumetric flow rate to pressure drop and pipe dimensions?

   (a) Darcy's Law

   (b) Fick's Law

   (c) Hagen-Poiseuille Equation

   (d) Fourier's Law

6. What is the primary role of Computational Fluid Dynamics (CFD) in relation to the Navier-Stokes equations?

   (a) Simplify the equations

   (b) Provide exact analytical solutions

   (c) Numerically solve the Navier-Stokes equations for complex flow problems

   (d) Eliminate the need for experimental validation

7. Which numerical method is commonly used to solve the Navier-Stokes equations on discretized domains?

(a) Finite Difference Method

(b) Finite Volume Method

(c) Variational Method

(d) Finite Element Method

Answers:
1. **B: Navier-Stokes equations** The Navier-Stokes equations form the fundamental basis for describing fluid motion, incorporating viscosity and pressure forces.

2. **C: Conservation of mass and momentum** The derivation of the Navier-Stokes equations is based on applying Newton's second law to fluid elements, considering mass and momentum conservation.

3. **B:** $\nabla \cdot \mathbf{u} = 0$ The continuity equation for incompressible fluids states that the divergence of the velocity field is zero, ensuring mass conservation.

4. **C:** $\mu \nabla^2 \mathbf{u}$ The term $\mu \nabla^2 \mathbf{u}$ in the Navier-Stokes equations accounts for viscous effects, representing internal friction within the fluid.

5. **C: Hagen-Poiseuille Equation** Hagen-Poiseuille equation relates the volumetric flow rate through a pipe to the pressure drop, viscosity, and dimensions of the pipe, derived from solutions to simplified Navier-Stokes equations under laminar flow conditions.

6. **C: Numerically solve the Navier-Stokes equations for complex flow problems** CFD involves using numerical algorithms to find approximate solutions to the Navier-Stokes equations, enabling analysis of fluid behavior in complex systems.

7. **B: Finite Volume Method** The finite volume method is widely used in fluid dynamics to solve differential equations like the Navier-Stokes by discretizing the domain into control volumes.

Chapter 2

Reynolds-Averaged Navier-Stokes (RANS)

Introduction to RANS Equations

The Reynolds-Averaged Navier-Stokes (RANS) equations are pivotal in the study of turbulent flows. They are derived by decomposing instantaneous quantities into mean and fluctuating components, facilitating the modeling of time-averaged effects of turbulence on mean flow fields.

Derivation of RANS Equations

Beginning with the decomposition of the velocity field, the velocity components **u**, **v**, and **w** are expressed as:

$$\mathbf{u} = \overline{\mathbf{u}} + \mathbf{u}'$$

$$\mathbf{v} = \overline{\mathbf{v}} + \mathbf{v}'$$

$$\mathbf{w} = \overline{\mathbf{w}} + \mathbf{w}'$$

where $\overline{\mathbf{u}}$ represents the mean velocity and $\mathbf{u}'$ the fluctuating component. Substituting these decompositions into the Navier-Stokes equations and applying time averaging lead to the RANS equations, capturing averaged effects while introducing the Reynolds stress tensor $\overline{\mathbf{u}'\mathbf{u}'}$.

Reynolds Stress Tensor

The Reynolds stress tensor embodies the influence of velocity fluctuations:

$$\overline{u'_i u'_j} = \nu_t \left(\partial_j \overline{u}_i + \partial_i \overline{u}_j\right) - \frac{2}{3} k \delta_{ij}$$

where ν_t denotes the eddy viscosity, k the turbulent kinetic energy, and δ_{ij} the Kronecker delta. This tensor is a fundamental component complicating the direct closure of the RANS equations.

Closure Models

Turbulence modeling involves devising closure models to approximate the Reynolds stress tensor. The most prevalent models utilize eddy viscosity hypotheses:

1 The Boussinesq Hypothesis

The Boussinesq hypothesis simplifies the stress-strain relationship by postulating that the eddy viscosity ν_t is isotropic and similar to molecular viscosity via:

$$\nu_t = C_\mu \frac{k^2}{\epsilon}$$

where C_μ is a model constant, k denotes turbulent kinetic energy, and ϵ represents its dissipation rate.

2 k-epsilon Model

The k-epsilon model, a two-equation model, extends this principle by introducing transport equations for k and ϵ:

$$\frac{\partial k}{\partial t} + \overline{u}_j \frac{\partial k}{\partial x_j} = \tau_{ij} \frac{\partial \overline{u}_i}{\partial x_j} - \epsilon + \frac{\partial}{\partial x_j}\left[\left(\nu + \frac{\nu_t}{\sigma_k}\right)\frac{\partial k}{\partial x_j}\right]$$

$$\frac{\partial \epsilon}{\partial t} + \overline{u}_j \frac{\partial \epsilon}{\partial x_j} = C_{1\epsilon}\frac{\epsilon}{k}\tau_{ij}\frac{\partial \overline{u}_i}{\partial x_j} - C_{2\epsilon}\frac{\epsilon^2}{k} + \frac{\partial}{\partial x_j}\left[\left(\nu + \frac{\nu_t}{\sigma_\epsilon}\right)\frac{\partial \epsilon}{\partial x_j}\right]$$

where σ_k and σ_ϵ are turbulent Prandtl numbers, with $C_{1\epsilon}$ and $C_{2\epsilon}$ as empirical constants.

Simulation Techniques for RANS

Simulation of RANS models integrates discretization methods and numerical solving strategies:

1 Finite Volume Method (FVM)

The FVM is crucial for solving RANS equations by discretizing the computational domain into control volumes where governing equations are applied in an integral form. It offers flexibility in mesh generation and conservation properties essential for accurate turbulence modeling.

2 Numerical Solvers

Iterative solvers such as SIMPLE (Semi-Implicit Method for Pressure Linked Equations) and PISO (Pressure-Implicit with Splitting of Operators) are implemented to resolve the coupled momentum and pressure fields:
   - SIMPLE algorithm facilitates the iterative correction of pressure and velocity fields to maintain continuity. - PISO extends this by enhancing convergence speed via improved pressure-velocity coupling.

Applications in Engineering

RANS models find extensive applications in numerous engineering fields, notably:

1 Aerodynamics

In aerodynamics, the prediction of lift and drag forces around airfoils using RANS models is instrumental in the design of efficient wings and fuselage shapes.

2 Hydraulic Systems

In hydraulic systems, RANS simulations provide insights into flow resistance and energy losses, optimizing pipeline and channel design.

The RANS framework underpins the feasibility of simulating complex turbulent flows with acceptable computational effort, providing the backbone for practical applications across engineering disciplines.

Python Code Snippet

Below is a Python code snippet that encompasses the core computational elements of the chapter's exploration of RANS equations, Reynolds stress tensor calculation, Boussinesq hypothesis application, k-epsilon model implementation, and numerical methods including SIMPLE and PISO algorithms.

```
import numpy as np

def decompose_velocity(velocity):
    '''
    Decomposes a velocity field into mean and fluctuating
    ↪ components.
    :param velocity: Instantaneous velocity value.
    :return: Mean velocity, fluctuating velocity component.
    '''
    mean_velocity = np.mean(velocity)
    fluctuating_component = velocity - mean_velocity
    return mean_velocity, fluctuating_component

def reynolds_stress_tensor(u_prime, v_prime, nu_t, dudx, dvdy, k,
↪ delta):
    '''
    Computes the Reynolds stress tensor component.
    :param u_prime: Fluctuating component of velocity u.
    :param v_prime: Fluctuating component of velocity v.
    :param nu_t: Eddy viscosity.
    :param dudx: Derivative of mean u velocity.
    :param dvdy: Derivative of mean v velocity.
    :param k: Turbulent kinetic energy.
    :param delta: Kronecker delta function.
    :return: Reynolds stress tensor component.
    '''
    return nu_t * (dudx + dvdy) - (2 / 3) * k * delta

def boussinesq_hypothesis(C_mu, k, epsilon):
    '''
    Applies the Boussinesq hypothesis to compute eddy viscosity.
    :param C_mu: Model constant.
    :param k: Turbulent kinetic energy.
    :param epsilon: Dissipation rate of turbulent kinetic energy.
    :return: Eddy viscosity.
```

```python
    '''
    return C_mu * (k ** 2) / epsilon

def k_epsilon_model(C1, C2, sigma_k, sigma_epsilon, nu, nu_t,
↪   tau_ij, dudx):
    '''
    Performs k-epsilon model calculations.
    :param C1: Constant C1epsilon.
    :param C2: Constant C2epsilon.
    :param sigma_k: Turbulent Prandtl number for k.
    :param sigma_epsilon: Turbulent Prandtl number for epsilon.
    :param nu: Molecular viscosity.
    :param nu_t: Eddy viscosity.
    :param tau_ij: Stress tensor component.
    :param dudx: Derivative of mean velocity u with respect to x.
    :return: Calculated k and epsilon values.
    '''
    dk_dt = (tau_ij * dudx) - epsilon + nu * (nu_t / sigma_k)
    d_epsilon_dt = C1 * (epsilon / k) * (tau_ij * dudx) - C2 *
↪   (epsilon ** 2 / k)
    return dk_dt, d_epsilon_dt

def finite_volume_method(volume, flow_field, boundary_conditions):
    '''
    Implements finite volume method for RANS equation
↪       discretization.
    :param volume: Control volume.
    :param flow_field: Discretized flow field.
    :param boundary_conditions: Boundary conditions for solver.
    :return: Resulting discretized field.
    '''
    # Placeholder operation representing discretization
    return flow_field * volume * boundary_conditions

def simple_algorithm(pressure_field, velocity_field):
    '''
    Applies the SIMPLE algorithm for pressure-velocity coupling.
    :param pressure_field: Discretized pressure field.
    :param velocity_field: Discretized velocity field.
    :return: Corrected pressure and velocity fields.
    '''
    # Simplified iterative correction step
    corrected_pressure = pressure_field + 0.5 * (velocity_field -
↪       pressure_field)
    corrected_velocity = velocity_field + 0.5 * (pressure_field -
↪       velocity_field)
    return corrected_pressure, corrected_velocity

def piso_algorithm(corrected_pressure, velocity_corr, iterations=1):
    '''
    Applies the PISO algorithm for enhanced convergence.
    :param corrected_pressure: Initial corrected pressure.
    :param velocity_corr: Initial corrected velocity.
```

```
:param iterations: Number of iterations for correction.
:return: Final pressure and velocity corrections.
'''
    for _ in range(iterations):
        corrected_pressure += 0.1 * (velocity_corr -
        ↪ corrected_pressure)
        velocity_corr += 0.1 * (corrected_pressure - velocity_corr)
    return corrected_pressure, velocity_corr

# Example of use
velocity_data = np.array([5, 15, 10, 8, 12])
u_mean, u_prime = decompose_velocity(velocity_data)
reynolds_stress = reynolds_stress_tensor(u_prime, u_prime, 0.07,
↪ 0.1, 0.2, 0.5, 1)
nu_t = boussinesq_hypothesis(0.09, 1, 1)
k_dt, epsilon_dt = k_epsilon_model(1.44, 1.92, 1, 1.3, 0.001, nu_t,
↪ reynolds_stress, 0.1)
discretized_field = finite_volume_method(0.5, np.array([1, 2, 3]),
↪ 0.9)
corrected_pressure, corrected_velocity =
↪ simple_algorithm(np.array([101, 99]), np.array([10, 12]))
final_pressure, final_velocity = piso_algorithm(corrected_pressure,
↪ corrected_velocity, 3)

print("Mean Velocity:", u_mean)
print("Reynolds Stress:", reynolds_stress)
print("Eddy Viscosity:", nu_t)
print("Discretized Field:", discretized_field)
print("Corrected Pressure/Velocity (SIMPLE):", corrected_pressure,
↪ corrected_velocity)
print("Final Pressure/Velocity (PISO):", final_pressure,
↪ final_velocity)
```

This code defines several key functions necessary for implementing the chapter's topics:

- `decompose_velocity` function calculates the mean and fluctuating components of a velocity field.

- `reynolds_stress_tensor` computes components of the Reynolds stress tensor using velocity fluctuations.

- `boussinesq_hypothesis` implements the Boussinesq hypothesis to determine eddy viscosity.

- `k_epsilon_model` calculates changes to turbulent kinetic energy and its dissipation rate.

- `finite_volume_method` discretizes RANS equations over a control volume using boundary conditions.

- `simple_algorithm` and `piso_algorithm` provide numerical methods for solving pressure and velocity coupling in fluid flows.

The final block of code provides examples of computing these elements using dummy data.

Multiple Choice Questions

1. What is the primary purpose of the Reynolds-Averaged Navier-Stokes (RANS) equations in fluid dynamics?

    (a) To solve the full Navier-Stokes equations directly

    (b) To analyze laminar flow behaviors

    (c) To model the mean effects of turbulence on flow fields

    (d) To describe molecular interactions in fluids

2. In the context of RANS, which term represents the fluctuating component of the velocity field?

    (a) $\bar{u}$

    (b) u'

    (c) v

    (d) $\overline{u'}$

3. Which mathematical model is commonly used to describe the Reynolds stress tensor within the RANS framework?

    (a) The Boussinesq Hypothesis

    (b) Newton's Second Law

    (c) Bernoulli's Principle

    (d) Pascal's Law

4. What is the primary goal of a closure model in RANS equations?

    (a) To derive new equations

    (b) To eliminate pressure terms

    (c) To approximate unresolved turbulent effects

    (d) To simplify computational algorithms

5. What do the empirical constants $C_{1\epsilon}$ and $C_{2\epsilon}$ relate to within the `k-epsilon` turbulence model?

    (a) Molecular diffusion rates

    (b) Boundary layer thickness

    (c) Energy transfer rates in turbulence

    (d) Compressibility effects in fluids

6. Which numerical method is frequently utilized to discretize RANS equations for computational simulations?

    (a) Finite Element Method (FEM)

    (b) Finite Volume Method (FVM)

    (c) Lattice Boltzmann Method (LBM)

    (d) Spectral Method

7. In what area of engineering are RANS models predominantly used to optimize the design of components?

    (a) Chemical processing

    (b) Structural engineering

    (c) Aerodynamics and hydraulic systems

    (d) Electrical systems

Answers:

1. **C: To model the mean effects of turbulence on flow fields** RANS equations are formulated to capture the average effects of turbulence on the flow, particularly in situations where capturing detail of instantaneous fluctuations is not feasible.

2. **B: u′** The notation **u′** denotes the fluctuating component of the velocity, essential for representing deviations from mean flow in turbulent dynamics.

3. **A: The Boussinesq Hypothesis** The Boussinesq Hypothesis provides a practical approach to modeling Reynolds stresses, assuming they can be approximated using an eddy viscosity model similar to molecular viscosity concepts.

4. **C: To approximate unresolved turbulent effects** Closure models are essential in RANS to predict the effects of turbulence that the equations themselves do not inherently resolve, allowing for more practical computation.

5. **C: Energy transfer rates in turbulence** The constants $C_{1\epsilon}$ and $C_{2\epsilon}$ are related to the production and dissipation terms

in the turbulence energy equations, reflecting the rates of energy transfer in the turbulent regime.

6. **B: Finite Volume Method (FVM)** The Finite Volume Method is widely used in computational fluid dynamics, including for RANS equations, because it ensures conservative properties and accommodates complex geometries efficiently.

7. **C: Aerodynamics and hydraulic systems** RANS models are extensively applied in aerodynamics for optimizing shapes and designs to improve aerodynamic performance, and in hydraulic systems for efficient fluid flow management.

Chapter 3

Turbulence Kinetic Energy Equation

Introduction to Turbulence Kinetic Energy (TKE)

The turbulence kinetic energy (TKE) equation serves as a fundamental component in the analysis of turbulent flows, addressing the distribution and transformation of energy within the turbulent spectrum. TKE, denoted by k, represents the kinetic energy per unit mass associated with eddies in a turbulent flow. The equation is derived from the Navier-Stokes equations, incorporating both production and dissipation terms.

Derivation of the TKE Equation

Initiating with the Navier-Stokes equations for incompressible flow, the turbulence kinetic energy equation is obtained by multiplying the momentum equations by the respective fluctuating velocity components and performing Reynolds averaging. The resultant form of the TKE equation is given by:

$$\frac{\partial k}{\partial t} + \bar{u}_j \frac{\partial k}{\partial x_j} = P_k - \epsilon + \frac{\partial}{\partial x_j} \left(\nu \frac{\partial k}{\partial x_j} + \frac{1}{2} \overline{u'_i u'_i u'_j} \right)$$

where P_k signifies the production term, ϵ the dissipation rate, and the final term is the diffusion by both molecular and turbulent processes.

Production of Turbulence Kinetic Energy

The production term P_k describes the conversion of mean flow kinetic energy into turbulence kinetic energy, typically through the mechanism of shear. This term is primarily governed by Reynolds stresses interacting with mean velocity gradients and is mathematically expressed as:

$$P_k = -\overline{u'_i u'_j} \frac{\partial \overline{u}_i}{\partial x_j}$$

The negative sign indicates the energy transfer from the mean flow to the turbulence. This production is most pronounced in regions of high velocity gradients such as boundary layers and shear layers.

Dissipation of Turbulence Kinetic Energy

The dissipation rate ϵ quantifies the energy lost from turbulence kinetic energy due to viscous effects at small scales. The dissipative nature of turbulence ensures that energy cascades from larger to smaller eddies until converted into thermal energy through molecular viscosity. The dissipation rate is defined in terms of velocity gradients and kinematic viscosity ν:

$$\epsilon = \nu \overline{\left(\frac{\partial u'_i}{\partial x_j}\right)^2}$$

This relationship underlines the inherent role of viscosity in diminishing turbulence kinetic energy at the microscale.

Equilibrium and Imbalance Dynamics

In practical applications, an equilibrium state in TKE may often be presumed where production balances dissipation, particularly in fully developed turbulent flows. However, departures from equilibrium can result in complex dynamics that drive turbulent transport processes. In engineering applications, maintaining control over these dynamics is crucial for optimizing performance and safety.

Numerical Implementation of TKE Equation

Simulation of flows involving TKE necessitates discretization techniques such as the Finite Volume Method (FVM), which conserves fluxes at control volume interfaces. Computational approaches often pair the TKE equation with a dissipation equation, such as the `k-epsilon` model, providing closure for Reynolds-Averaged Navier-Stokes equations:

$$\frac{\partial k}{\partial t} + \bar{u}_j \frac{\partial k}{\partial x_j} = \nu_t \frac{\partial^2 k}{\partial x_j^2} + P_k - \epsilon$$

The implemented algorithms ensure that the intricate balance of energy production, dissipation, and transport is captured with fidelity.

Applications in Engineering System Analyses

The incorporation of turbulence kinetic energy equations into engineering design and analysis enhances the understanding of turbulent scaling effects and energy distribution across varied flow regimes. In aerodynamics, accurate prediction of turbulence-driven boundary layer properties hinges upon proficient modeling of TKE dynamics. Similarly, in hydraulic systems, TKE insights assist in efficiently managing flow resistance and optimizing energy throughput in pipe networks and flow channels.

Python Code Snippet

Below is a Python code snippet that demonstrates the core computational elements of turbulence kinetic energy (TKE) equation derivation, production, dissipation, and its numerical implementation using basic discretization techniques in computational fluid dynamics (CFD).

```
import numpy as np

def tke_production(reynolds_stress, mean_velocity_gradient):
```

```python
    '''
    Calculate the production of turbulence kinetic energy.
    :param reynolds_stress: Array of Reynolds stresses.
    :param mean_velocity_gradient: Gradient of mean velocity.
    :return: Production of TKE.
    '''
    return -np.sum(reynolds_stress * mean_velocity_gradient)

def tke_dissipation(velocity_gradients, kinematic_viscosity):
    '''
    Calculate the dissipation of turbulence kinetic energy.
    :param velocity_gradients: Array of velocity gradients.
    :param kinematic_viscosity: Kinematic viscosity of the fluid.
    :return: Dissipation rate of TKE.
    '''
    return kinematic_viscosity * np.sum(velocity_gradients**2)

def tke_equation(tke, mean_velocity, production, dissipation,
   ↪ turbulent_diffusion, time_step, spatial_step):
    '''
    Simulate the time evolution of turbulence kinetic energy.
    :param tke: Initial TKE.
    :param mean_velocity: Mean flow velocity array.
    :param production: TKE production term.
    :param dissipation: TKE dissipation rate.
    :param turbulent_diffusion: Turbulent diffusion term.
    :param time_step: Simulation time step.
    :param spatial_step: Simulation spatial discretization step.
    :return: Updated TKE.
    '''
    convective_term = np.dot(mean_velocity, np.gradient(tke,
       ↪ spatial_step))
    diffusive_term = np.gradient(turbulent_diffusion, spatial_step)

    tke_next = tke + time_step * (production - dissipation -
       ↪ convective_term + diffusive_term)

    return tke_next

# Define system parameters for simulation
mean_velocity = np.array([3.0, 2.0, 1.0])                    # Example
   ↪ mean velocities
reynolds_stress = np.array([0.1, 0.1, 0.1])                  # Example
   ↪ Reynolds stresses
velocity_gradient = np.array([0.5, 0.0, -0.5])               # Velocity
   ↪ gradient
kinematic_viscosity = 1e-5                                   #
   ↪ Kinematic viscosity
tke_initial = 1.0                                            # Initial
   ↪ TKE
turbulent_diffusion = 0.001                                  #
   ↪ Turbulent diffusion term
```

```
time_step = 0.01                                          # Time
↪   step for simulation
spatial_step = 1.0                                        # Spatial
↪   step for simulation

# Calculate production and dissipation
production = tke_production(reynolds_stress, velocity_gradient)
dissipation = tke_dissipation(velocity_gradient,
↪   kinematic_viscosity)

# Simulate TKE evolution for one step
tke_next = tke_equation(tke_initial, mean_velocity, production,
↪   dissipation,
                        turbulent_diffusion, time_step,
                        ↪   spatial_step)

print("Initial TKE:", tke_initial)
print("Production:", production)
print("Dissipation:", dissipation)
print("Next TKE:", tke_next)
```

This code defines several key functions necessary for the computation and simulation of turbulence kinetic energy dynamics:

- `tke_production` computes the production of turbulence kinetic energy based on Reynolds stresses and velocity gradients.

- `tke_dissipation` calculates the energy dissipation rate using velocity gradients and fluid viscosity.

- `tke_equation` simulates the time evolution of TKE by incorporating production, dissipation, convection, and diffusion terms.

The code demonstrates the evolution of TKE over a single simulation step using dummy parameters for clarity.

Multiple Choice Questions

1. The turbulence kinetic energy (TKE) equation is primarily derived from which set of equations?

    (a) Euler Equations

    (b) Navier-Stokes Equations

    (c) Bernoulli's Equation

(d) Vorticity Transport Equation

2. In the TKE equation, what does the term ϵ represent?

    (a) Turbulence production rate
    (b) Turbulence dissipation rate
    (c) Turbulence diffusion rate
    (d) Turbulence convection rate

3. Which of the following terms in the TKE equation is responsible for energy transfer from the mean flow to turbulence?

    (a) Diffusion term
    (b) Production term
    (c) Dissipation term
    (d) Convection term

4. The production of turbulence kinetic energy is most pronounced in which regions?

    (a) Regions of low velocity gradients
    (b) Irrotational regions
    (c) Regions of high velocity gradients
    (d) Laminar flow regions

5. In computational simulations, which method is commonly used to discretize the TKE equation?

    (a) Finite Difference Method
    (b) Finite Element Method
    (c) Finite Volume Method
    (d) Spectral Method

6. The equation used in conjunction with the TKE equation to provide closure for the RANS equations often involves what additional variable?

    (a) Density
    (b) Enthalpy
    (c) Dissipation
    (d) Temperature

7. What energy cascade mechanism in turbulence is described by the TKE dissipation term ϵ?

   (a) Conversion of thermal energy to kinetic energy
   (b) Transfer of energy from smaller to larger eddies
   (c) Transfer of energy from larger to smaller eddies
   (d) Conservation of kinetic energy in turbulent flow

Answers:
1. **B: Navier-Stokes Equations** The TKE equation is derived from the Navier-Stokes equations, which describe how the velocity field of a fluid evolves, incorporating various terms like production and dissipation relevant to turbulence.

2. **B: Turbulence dissipation rate** In the TKE equation, ϵ represents the dissipation rate, quantifying the rate at which turbulence kinetic energy is converted to thermal energy through viscous effects.

3. **B: Production term** The production term P_k in the TKE equation represents the transfer of kinetic energy from the mean flow into turbulence due to velocity gradients, primarily through shear mechanisms.

4. **C: Regions of high velocity gradients** Turbulence production is most pronounced in regions of high velocity gradients, such as boundary layers and shear layers, where the flow dynamics promote the transfer of energy into turbulent forms.

5. **C: Finite Volume Method** The Finite Volume Method is commonly used for discretizing the TKE equation in computational simulations because it conserves fluxes at the control volume interfaces, accounting for convective and diffusive transport of turbulence.

6. **C: Dissipation** The TKE equation is commonly used with a dissipation equation, such as those present in turbulence models like k-epsilon, to close the Reynolds-Averaged Navier-Stokes (RANS) equations.

7. **C: Transfer of energy from larger to smaller eddies** The dissipation term ϵ in the TKE equation describes the energy cascade in turbulence, where energy is transferred from larger to smaller eddies until it is dissipated as heat due to viscosity.

Chapter 4

Bernoulli's Equation - Extended Application

Foundation of Bernoulli's Equation

Bernoulli's equation represents a fundamental principle in fluid dynamics, describing the conservation of energy in a streamline flow. The general form of Bernoulli's equation for steady, incompressible flow along a streamline is given by:

$$P + \frac{1}{2}\rho v^2 + \rho g h = \text{constant}$$

where P is the static pressure, ρ is the fluid density, v is the flow velocity, and gh is the gravitational potential energy per unit volume. This equation is central to analyzing various fluid systems by correlating pressure, velocity, and elevation.

Extended Bernoulli's Equation in Varying Cross-Sections

In real-world applications, fluid flows through systems with varying cross-sections, necessitating modifications to the classical Bernoulli equation to account for additional energy considerations. Specifically, energy losses due to friction and changes in cross-sectional area must be considered when analyzing such scenarios.

For a segment of pipe with varying diameters, incorporating head loss terms h_f due to pipe friction, Bernoulli's equation can be written as:

$$P_1 + \frac{1}{2}\rho v_1^2 + \rho g h_1 = P_2 + \frac{1}{2}\rho v_2^2 + \rho g h_2 + \rho g h_f$$

where subscript 1 indicates the upstream section and subscript 2 the downstream section. The term $\rho g h_f$ accounts for energy losses primarily due to viscous effects and can be quantified using empirical correlations or Moody's diagram in pipe flow analysis.

Unsteady Flow Considerations

Analyzing unsteady flow extends Bernoulli's implications to account for temporal variations in flow properties. In unsteady conditions, the flow parameters such as pressure and velocity vary with time, necessitating a time-dependent analysis framework.

For unsteady flow, the temporal derivative introduces an additional term into the Bernoulli equation. Thus, the extended form for an unsteady flow scenario becomes:

$$\frac{\partial}{\partial t}\left(P + \frac{1}{2}\rho v^2 + \rho g h\right) = -\rho \frac{\partial \Phi}{\partial t}$$

where Φ represents the velocity potential, and its temporal derivative reflects changes in the energy field with respect to time.

Practical Applications in Hydraulic Systems

In hydraulic systems with complex geometries and unsteady conditions, Bernoulli's equation must be adeptly applied to predict system performance. For example, in pump and turbine design, Bernoulli's principle aids in evaluating energy conversion processes by linking pressure and velocity changes to mechanical power output.

When applied to flow measurement devices such as Venturi meters or flow nozzles, the differential pressure interpreted through Bernoulli's framework offers insight into volumetric flow rates. The pressure difference, due to changes in cross-sections, correlates with velocity changes through:

$$\Delta P = \frac{\rho}{2}(v_2^2 - v_1^2)$$

This relationship forms the basis of converting differential pressure readings into accurate flow measurements, a critical task in process and chemical engineering sectors.

Mathematical Modelling in Computational Fluid Dynamics

In Computational Fluid Dynamics (CFD), the extended Bernoulli equation provides constraints and verification checks during the simulation of fluid flow systems. Implementing Bernoulli's augmented terms into the Navier-Stokes equations facilitates the numerical assessment of fluid flow scenarios through grid-based computations.

Advanced CFD applications leverage the Bernoulli principle to validate simulation results. By ensuring that computed pressure and velocity distributions adhere to Bernoulli's predictions on a macroscopic level, model accuracy in recreating physical flow phenomena is ascertained.

In summary, while the classical Bernoulli equation underpins energy conservation in fluid flow, its extended forms, incorporating frictional and temporal considerations, enhance the robustness and applicability of this principle in engineering contexts. Through careful adaptation to variable cross-sections and unsteady conditions, Bernoulli's equation remains an indispensable analytical tool in the realm of fluid dynamics.

Python Code Snippet

Below is a Python code snippet that illustrates key computations based on Bernoulli's equation and its extended applications, including handling scenarios with varying cross-sections, unsteady flows, and frictional losses.

```
import numpy as np
def bernoulli_steady(P1, v1, h1, P2, v2, h2, rho):
    '''
```

```python
    Calculate the Bernoulli's equation term for steady flows with
    ↪ head loss.
    :param P1: Upstream pressure.
    :param v1: Upstream velocity.
    :param h1: Upstream elevation.
    :param P2: Downstream pressure.
    :param v2: Downstream velocity.
    :param h2: Downstream elevation.
    :param rho: Fluid density.
    :return: Needed head loss for balance.
    '''
    return (P1 + 0.5 * rho * v1**2 + rho * 9.81 * h1) - (P2 + 0.5 *
    ↪ rho * v2**2 + rho * 9.81 * h2)

def head_loss(pipe_length, diameter, velocity, kinematic_viscosity,
↪ roughness):
    '''
    Calculate head loss using Darcy-Weisbach equation.
    :param pipe_length: Length of pipe.
    :param diameter: Diameter of pipe.
    :param velocity: Fluid velocity.
    :param kinematic_viscosity: Kinematic viscosity of fluid.
    :param roughness: Pipe roughness.
    :return: Head loss due to friction.
    '''
    reynolds_number = (velocity * diameter) / kinematic_viscosity
    friction_factor = 0.02   # Assume smooth pipe approximation,
    ↪ replace with Moody chart value
    return (friction_factor * (pipe_length / diameter) *
    ↪ (velocity**2 / (2 * 9.81)))

def bernoulli_unsteady(P, v, h, rho, dPhi_dt):
    '''
    Extended Bernoulli's equation for unsteady flow.
    :param P: Pressure.
    :param v: Velocity.
    :param h: Elevation.
    :param rho: Fluid density.
    :param dPhi_dt: Change in velocity potential with time.
    :return: Time-dependent Bernoulli's result.
    '''
    return -(P + 0.5 * rho * v**2 + rho * 9.81 * h) - rho * dPhi_dt

def flow_measure_differential_pressure(rho, v1, v2):
    '''
    Calculate pressure difference for flow measurement devices.
    :param rho: Fluid density.
    :param v1: Velocity at point 1.
    :param v2: Velocity at point 2.
    :return: Differential pressure.
    '''
    return 0.5 * rho * (v2**2 - v1**2)
```

```
# Sample parameters for demonstration
P1, v1, h1 = 101325, 3, 0   # Upstream conditions
P2, v2, h2 = 95000, 5, 2    # Downstream conditions
rho = 1000   # Density of water in kg/m^3
pipe_length, diameter = 30, 0.5   # Pipe attributes
kinematic_viscosity = 1.004e-6   # Water's kinematic viscosity in
    m^2/s
roughness = 0.0001   # Roughness of pipe
dPhi_dt = 0.02   # Change in velocity potential with time

# Calculations
steady_balance = bernoulli_steady(P1, v1, h1, P2, v2, h2, rho)
head_loss_val = head_loss(pipe_length, diameter, v1,
    kinematic_viscosity, roughness)
unsteady_result = bernoulli_unsteady(P1, v1, h1, rho, dPhi_dt)
differential_pressure = flow_measure_differential_pressure(rho, v1,
    v2)

print("Steady Bernoulli Balance:", steady_balance)
print("Head Loss:", head_loss_val)
print("Unsteady Flow Result:", unsteady_result)
print("Differential Pressure for Flow Measurement:",
    differential_pressure)
```

This code provides several functions to explore different aspects of Bernoulli's equation within complex engineering contexts:

- `bernoulli_steady`: Computes the balance required in steady flows incorporating head loss for varying pressure, velocity, and height.

- `head_loss`: Uses the Darcy-Weisbach equation to calculate frictional losses along a pipeline.

- `bernoulli_unsteady`: Adapts Bernoulli's principle to account for temporal variations within unsteady flows.

- `flow_measure_differential_pressure`: Evaluates differential pressures in flow measurement scenarios given velocity changes.

The snippet exemplifies practical applications using example parameters, with core outputs provided for each scenario.

Multiple Choice Questions

1. Which term in Bernoulli's Equation represents the kinetic energy per unit volume of the fluid?

(a) P

(b) $\frac{1}{2}\rho v^2$

(c) $\rho g h$

(d) $\rho g h_f$

2. In a fluid system with varying cross-sections, which factor primarily contributes to head loss?

    (a) Changes in elevation

    (b) Friction due to pipe walls

    (c) Variations in fluid pressure

    (d) Fluid compressibility

3. How does the extended Bernoulli's Equation modify the classical form to account for unsteady flow conditions?

    (a) By adding a velocity potential term

    (b) By eliminating the gravitational potential energy term

    (c) By introducing a temporal derivative term

    (d) By changing the flow velocity term

4. For flow measurement devices such as Venturi meters, the Bernoulli equation is used to relate velocity changes to:

    (a) Temperature changes

    (b) Pressure differences

    (c) Density variations

    (d) Flow direction

5. Which parameter in the Bernoulli equation accounts for energy losses due to viscous effects?

    (a) ρv^2

    (b) $\rho g h$

    (c) $\rho g h_f$

    (d) Φ

6. How is the concept of Bernoulli's principle applied within Computational Fluid Dynamics (CFD)?

    (a) As a primary equation for solving fluid flow problems

(b) For constraint checking and result verification

(c) To simplify Navier-Stokes equations

(d) To automatically calculate fluid viscosity

7. When analyzing pump and turbine systems, Bernoulli's equation helps in understanding:

(a) Fluid condensation processes

(b) Energy conversion efficiency

(c) Electrical conductivity of fluids

(d) Turbulence intensities

Answers:

1. **B:** $\frac{1}{2}\rho v^2$ This term represents the kinetic energy per unit volume, where ρ is the fluid density and v is the flow velocity.

2. **B: Friction due to pipe walls** Frictional losses in pipes, which result in head loss, are primarily due to the interaction between the fluid and the pipe wall surface.

3. **C: By introducing a temporal derivative term** In unsteady flow, the Bernoulli equation is extended with a time-dependent term to account for temporal changes in flow properties.

4. **B: Pressure differences** Devices like Venturi meters measure flow velocity changes by correlating them with pressure differences, as predicted by the Bernoulli equation.

5. **C:** $\rho g h_f$ This term is included to account for head losses due to viscous effects, often described using empirical factors or frictional correlations.

6. **B: For constraint checking and result verification** In CFD, Bernoulli's equation is used to verify the accuracy of simulations, ensuring that the energy conservation principle is maintained.

7. **B: Energy conversion efficiency** Bernoulli's equation helps in evaluating how pressure and velocity changes relate to the energy exchanged in pumps and turbines, crucial for understanding their performance.

Chapter 5

Vorticity Transport Equation

Vorticity in Fluid Mechanics

Vorticity is a fundamental concept in fluid mechanics, representing the local spinning motion of a fluid. It is mathematically defined as the curl of the velocity field **v**, given by:

$$\boldsymbol{\omega} = \nabla \times \mathbf{v}$$

where $\boldsymbol{\omega}$ is the vorticity vector. In a three-dimensional Cartesian coordinate system, the components of the vorticity can be expressed as:

$$\omega_x = \frac{\partial v_z}{\partial y} - \frac{\partial v_y}{\partial z}, \quad \omega_y = \frac{\partial v_x}{\partial z} - \frac{\partial v_z}{\partial x}, \quad \omega_z = \frac{\partial v_y}{\partial x} - \frac{\partial v_x}{\partial y}$$

This vector quantity provides insight into the rotational characteristics of fluid motion, critical for the analysis of turbulent and vortex-dominated flows.

Derivation of the Vorticity Transport Equation

The vorticity transport equation is derived from the Navier-Stokes equations and provides a framework to analyze the evolution of vorticity within a fluid. Starting with the incompressible Navier-Stokes equation:

$$\rho\left(\frac{\partial \mathbf{v}}{\partial t} + (\mathbf{v}\cdot\nabla)\mathbf{v}\right) = -\nabla p + \mu\nabla^2\mathbf{v}$$

By taking the curl of the momentum equation, the pressure gradient term vanishes and the vorticity transport equation is obtained:

$$\frac{\partial \boldsymbol{\omega}}{\partial t} + (\mathbf{v}\cdot\nabla)\boldsymbol{\omega} = (\boldsymbol{\omega}\cdot\nabla)\mathbf{v} + \nu\nabla^2\boldsymbol{\omega}$$

where ν is the kinematic viscosity. The equation includes terms for vorticity advection, vortex stretching, and diffusion.

Vortex Dynamics and Stretching

The vortex stretching term $(\boldsymbol{\omega}\cdot\nabla)\mathbf{v}$ in the vorticity transport equation plays a pivotal role in vortex dynamics. Vortex stretching is a mechanism through which vorticity is intensified due to the elongation of vortex lines in a velocity gradient field. In an incompressible flow, this phenomenon contributes significantly to the energy cascade observed in turbulent flows.

The stretching term can be examined under specific configurations. For instance, in axisymmetric flows, the vortex stretching is directly associated with changes in radial velocity gradients, which alter vorticity magnitudes.

Implications for Turbulent Flows

In turbulent flow regimes, the dynamics of vorticity are crucial for predicting the formation and evolution of vortical structures. Understanding the interplay between vorticity advection, stretching, and diffusion facilitates the modeling of turbulent mixing and dissipation.

The diffusion term, $\nu\nabla^2\boldsymbol{\omega}$, acts to smooth out gradients in vorticity, counteracting the effects of vortex stretching and advection. Its role becomes especially significant at smaller scales where viscous effects dominate.

Computational simulations often employ the vorticity transport equation as a basis for capturing the intricate dynamics of turbulence. Numerically resolving the vorticity field enables the visualization and study of coherent structures, such as vortices, within a turbulent flow.

Mathematical Modeling

In computational fluid dynamics (CFD), the integration of the vorticity transport equation requires sophisticated numerical techniques. Discretization methods such as finite difference, finite volume, or finite element methods facilitate solving the vorticity transport equation on computational grids.

Considerations for boundary conditions, especially in cases with wall-bounded flows, need careful treatment to accurately represent vorticity interactions at solid boundaries. The no-slip condition, for example, alters the vorticity distribution due to the viscous boundary layer effects.

Advanced turbulence models, including Large Eddy Simulation (LES) and Direct Numerical Simulation (DNS), incorporate the principles underlying vorticity transport to yield detailed predictions of turbulent flows, providing fidelity in capturing complex vortex dynamics inherent in engineering applications.

Python Code Snippet

Below is a Python code snippet that computes the key elements of the vorticity transport equation, providing utilities for simulation of vorticity dynamics, vortex stretching analysis, and numerical resolution scenarios using foundational computational physics methods.

```
import numpy as np

def curl(v):
    '''
    Calculate the vorticity vector as the curl of the velocity
     ↪ field.
    :param v: Velocity field represented as a 3D numpy array
    :return: Vorticity vector field
    '''
    dvz_dy = np.gradient(v[2], axis=1)
```

```python
        dvy_dz = np.gradient(v[1], axis=2)
        dvx_dz = np.gradient(v[0], axis=2)
        dvz_dx = np.gradient(v[2], axis=0)
        dvy_dx = np.gradient(v[1], axis=0)
        dvx_dy = np.gradient(v[0], axis=1)

        omega_x = dvz_dy - dvy_dz
        omega_y = dvx_dz - dvz_dx
        omega_z = dvy_dx - dvx_dy

        return np.array([omega_x, omega_y, omega_z])

def vorticity_transport(omega, v, nu, dt, grid_size):
    '''
    Compute the next timestep of the vorticity using the vorticity
    ↪   transport equation.
    :param omega: Current vorticity vector field
    :param v: Velocity field
    :param nu: Kinematic viscosity
    :param dt: Timestep size
    :param grid_size: Spatial discretization size
    :return: Updated vorticity field
    '''
    vx, vy, vz = np.gradient(v[0], grid_size), np.gradient(v[1],
    ↪   grid_size), np.gradient(v[2], grid_size)

    # Vorticity advection
    advect_omega = - (v[0] * np.gradient(omega[0], axis=0) +
                     v[1] * np.gradient(omega[0], axis=1) +
                     v[2] * np.gradient(omega[0], axis=2))

    # Vortex stretching
    stretch_omega = (omega[0] * vx[0] + omega[1] * vx[1] + omega[2]
    ↪   * vx[2])

    # Diffusion term
    diff_omega = nu * (np.gradient(np.gradient(omega[0], axis=0),
    ↪   axis=0) +
                      np.gradient(np.gradient(omega[0], axis=1),
                      ↪   axis=1) +
                      np.gradient(np.gradient(omega[0], axis=2),
                      ↪   axis=2))

    new_omega_x = omega[0] + dt * (advect_omega + stretch_omega +
    ↪   diff_omega)

    advect_omega = - (v[0] * np.gradient(omega[1], axis=0) +
                     v[1] * np.gradient(omega[1], axis=1) +
                     v[2] * np.gradient(omega[1], axis=2))

    stretch_omega = (omega[0] * vy[0] + omega[1] * vy[1] + omega[2]
    ↪   * vy[2])
```

```
diff_omega = nu * (np.gradient(np.gradient(omega[1], axis=0),
↪   axis=0) +
                  np.gradient(np.gradient(omega[1], axis=1),
↪       axis=1) +
                  np.gradient(np.gradient(omega[1], axis=2),
↪       axis=2))

new_omega_y = omega[1] + dt * (advect_omega + stretch_omega +
↪   diff_omega)

advect_omega = - (v[0] * np.gradient(omega[2], axis=0) +
                  v[1] * np.gradient(omega[2], axis=1) +
                  v[2] * np.gradient(omega[2], axis=2))

stretch_omega = (omega[0] * vz[0] + omega[1] * vz[1] + omega[2]
↪   * vz[2])

diff_omega = nu * (np.gradient(np.gradient(omega[2], axis=0),
↪   axis=0) +
                  np.gradient(np.gradient(omega[2], axis=1),
↪       axis=1) +
                  np.gradient(np.gradient(omega[2], axis=2),
↪       axis=2))

new_omega_z = omega[2] + dt * (advect_omega + stretch_omega +
↪   diff_omega)

return np.array([new_omega_x, new_omega_y, new_omega_z])

# Example of usage
grid_size = 0.01  # spatial discretization
dt = 0.01  # timestep
nu = 0.01  # kinematic viscosity

v = np.random.rand(3, 100, 100, 100)  # example velocity field
omega = curl(v)  # initial vorticity value

# Simulate one timestep of vorticity transport
new_omega = vorticity_transport(omega, v, nu, dt, grid_size)

print("Updated vorticity field after one timestep: ", new_omega)
```

This extended code defines the mathematical transformation and solution steps for the vorticity transport equation in a fluid dynamics context:

- curl function computes the vorticity from a given velocity field using finite differences.
- vorticity_transport calculates the evolution of vorticity

over a timestep, integrating advection, stretching, and diffusion based on the transport equation.

- Simulation showcases the update process within a 3D grid, reflecting realistic resolution of vorticity mechanics in CFD applications.

Utilizing these computational steps enables simulation of fluid dynamic behaviors critical to engineering and physical sciences fields, offering a foundation for further numerical analysis and visualization.

Multiple Choice Questions

1. What does vorticity represent in fluid mechanics?

    (a) The pressure gradient in a fluid

    (b) The local linear motion of a fluid

    (c) The local spinning motion of a fluid

    (d) The density fluctuation in a fluid

2. Which mathematical operation is used to define vorticity in terms of the velocity field?

    (a) Dot product

    (b) Cross product

    (c) Divergence

    (d) Curl

3. The vorticity transport equation is derived by taking the curl of which fundamental equation?

    (a) Continuity equation

    (b) Bernoulli's equation

    (c) Euler's equation

    (d) Navier-Stokes equation

4. In the vorticity transport equation, what process does the term $(\boldsymbol{\omega} \cdot \nabla)\mathbf{v}$ refer to?

    (a) Vorticity advection

(b) Vorticity diffusion

   (c) Vortex stretching

   (d) Pressure gradient force

5. What role does the diffusion term $\nu\nabla^2\omega$ play in the vorticity transport equation?

   (a) It increases the vorticity magnitude

   (b) It smooths out vorticity gradients

   (c) It enhances vortex stretching

   (d) It adds energy to the flow

6. How does the vorticity transport equation aid in understanding turbulent flows?

   (a) By analyzing laminar flow characteristics

   (b) By highlighting the role of pressure in fluid motion

   (c) By providing insights into the formation and evolution of vortical structures

   (d) By determining compressibility effects

7. Which computational techniques are mentioned for solving the vorticity transport equation in CFD?

   (a) Integral methods

   (b) Discrete vortex methods

   (c) Finite difference, finite volume, or finite element methods

   (d) Lattice Boltzmann methods

Answers:

1. **C: The local spinning motion of a fluid** Vorticity measures the tendency of fluid elements to spin about their own axes, which is crucial in analyzing rotational flow features.

2. **D: Curl** Vorticity is defined as the curl of the velocity field, resulting in a vector quantity that describes the rotation of fluid elements.

3. **D: Navier-Stokes equation** The vorticity transport equation is derived from the Navier-Stokes equation by taking its curl, which isolates rotational effects by removing the pressure term.

4. **C: Vortex stretching** The term $(\boldsymbol{\omega} \cdot \nabla)\mathbf{v}$ describes vortex stretching, a key mechanism for increasing vorticity in a velocity gradient field.

5. **B: It smooths out vorticity gradients** The diffusion term $\nu\nabla^2\boldsymbol{\omega}$ represents the viscous dissipation of vorticity, acting to even out sharp changes in vorticity.

6. **C: By providing insights into the formation and evolution of vortical structures** The key to understanding turbulence is analyzing how vortices form, interact, and dissipate, all of which are central to the vorticity transport mechanism.

7. **C: Finite difference, finite volume, or finite element methods** These numerical techniques discretize the equations onto a computational grid, enabling the simulation of complex flow dynamics, including vorticity transport.

Chapter 6

Potential Flow Theory

Governing Equations and Assumptions

Potential flow theory is grounded in the simplification of fluid motion equations, particularly under the assumptions of irrotational and incompressible fluid behavior. The velocity field $\mathbf{v}$ of a potential flow can be described as the gradient of a scalar potential function, ϕ, such that:

$$\mathbf{v} = \nabla \phi$$

Under the condition of incompressibility, the fluid density remains constant, resulting in the divergence-free condition for the velocity field:

$$\nabla \cdot \mathbf{v} = 0$$

Substituting the potential function into the above condition yields the Laplace equation for ϕ:

$$\nabla^2 \phi = 0$$

This equation is central to potential flow theory, providing the mathematical basis for flow potential problems.

Boundary Conditions

The solution to the Laplace equation requires the application of appropriate boundary conditions. In physical scenarios involving solid boundaries, the no-penetration condition is paramount. It ensures that the normal component of velocity at a solid surface is zero, represented as:

$$\frac{\partial \phi}{\partial n} = 0$$

where n denotes the outward normal at the boundary. For far-field conditions, the potential function often satisfies conditions such as uniform flow or vanishing disturbance, that dictate the behavior of ϕ as it approaches infinity.

Stream Function Formulation

In two-dimensional incompressible flow, the introduction of the stream function ψ complements the potential function framework. The relationship between the velocity components and the stream function is expressed as:

$$u = \frac{\partial \psi}{\partial y}, \quad v = -\frac{\partial \psi}{\partial x}$$

The continuity equation is inherently satisfied by these definitions, indicating the orthogonality of $\nabla \phi$ and $\nabla \psi$.

Complex Potential and Conformal Mapping

For two-dimensional potential flows, the complex potential $\Phi(z)$, combines the potential function ϕ and the stream function ψ into a single complex function:

$$\Phi(z) = \phi + i\psi$$

where $z = x + iy$ is the complex coordinate. The derivative of the complex potential with respect to z gives the complex velocity:

$$\frac{d\Phi}{dz} = u - iv$$

Conformal mapping techniques employ analytic functions to transform complex geometries into simpler ones while preserving angles, thus facilitating the solution of potential flow problems in complex domains.

Applications in External Flows

The applicability of potential flow theory extends to various classical problems, such as the flow around bodies. Examples include:

1 Flow Around a Cylinder

The potential function for the flow around a circular cylinder with radius R can be described using polar coordinates (r, θ):

$$\phi(r, \theta) = U \left(r + \frac{R^2}{r} \right) \cos \theta$$

The corresponding stream function takes the form:

$$\psi(r, \theta) = U \left(r - \frac{R^2}{r} \right) \sin \theta$$

These solutions adhere to the boundary conditions on the cylinder surface and asymptotic behavior at infinity.

2 Source and Sink Flows

Elementary potential flows involve point sources and sinks, characterized by radially symmetric ϕ. For a source located at the origin, the potential function is:

$$\phi(r) = \frac{Q}{2\pi} \ln r$$

where Q is the source strength. Particle paths around such singularities are radial lines emanating from the source.

Limitations and Idealizations

Despite its widespread use, potential flow theory exhibits inherent limitations due to its neglect of viscous effects. It aptly describes inviscid and non-separation flow regions but requires coupling with empirical or numerical techniques for accurate representation near boundary layers and regions with flow separation.

Python Code Snippet

Below is a Python code snippet that encapsulates the core computational elements of potential flow theory, including the Laplace equation solver for potential function, boundary conditions implementation, and visualization of flow around a cylinder.

```python
import numpy as np
import matplotlib.pyplot as plt
from matplotlib import cm

def solve_laplace(grid_size, boundary_conditions_function):
    '''
    Solve the Laplace equation using finite difference method.
    :param grid_size: Tuple indicating the size of the simulation
     ↪  grid.
    :param boundary_conditions_function: Function to apply boundary
     ↪  conditions.
    :return: Potential function phi on the grid.
    '''
    phi = np.zeros(grid_size)
    boundary_conditions_function(phi)

    # Iterate until convergence
    for _ in range(1000):
        phi_old = phi.copy()
        phi[1:-1, 1:-1] = 0.25 * (phi_old[:-2, 1:-1] + phi_old[2:,
         ↪  1:-1] +
                                  phi_old[1:-1, :-2] +
         ↪          phi_old[1:-1, 2:])
        boundary_conditions_function(phi)
        if np.max(np.abs(phi - phi_old)) < 1e-5:
            break

    return phi

def boundary_conditions(phi):
    '''
    Define boundary conditions for the potential flow.
    Reflective conditions at the edges of grid and specific
     ↪  condition for the obstacle.
    :param phi: Potential function grid.
    '''
    phi[0, :] = phi[-1, :]   # Top and bottom reflective boundaries
    phi[:, 0] = phi[:, -1]   # Left and right reflective boundaries
    # Insert specific obstacles or features here

def velocity_field(phi, dx):
    '''
    Calculate the velocity field from the potential function.
    :param phi: Potential function grid.
```

```python
    :param dx: Grid spacing.
    :return: Velocity field components u, v.
    '''
    u = np.zeros_like(phi)
    v = np.zeros_like(phi)

    u[1:-1, 1:-1] = (phi[1:-1, 2:] - phi[1:-1, :-2]) / (2 * dx)
    v[1:-1, 1:-1] = (phi[2:, 1:-1] - phi[:-2, 1:-1]) / (2 * dx)

    return u, v

def plot_flow(phi, u, v):
    '''
    Plot the potential function and velocity field lines.
    :param phi: Potential function.
    :param u: Velocities in x-direction.
    :param v: Velocities in y-direction.
    '''
    y, x = np.mgrid[0:phi.shape[0], 0:phi.shape[1]]
    fig, ax = plt.subplots(figsize=(8, 6))
    ax.contourf(x, y, phi, 50, cmap=cm.viridis)
    ax.streamplot(x, y, u, v, color='k', linewidth=0.5)
    plt.title('Potential Flow with Streamlines')
    plt.xlabel('x')
    plt.ylabel('y')
    plt.show()

# Parameters for the simulation
grid_size = (50, 50)
dx = 1.0

# Solving the potential flow problem
phi = solve_laplace(grid_size, boundary_conditions)
u, v = velocity_field(phi, dx)

# Plotting the results
plot_flow(phi, u, v)
```

This code defines several functions necessary for the implementation of potential flow simulations:

- `solve_laplace` solves the Laplace equation for a given grid, ensuring convergence through iterative finite difference approximations.

- `boundary_conditions` applies necessary conditions for reflective or specific boundary dynamics in the simulation grid.

- `velocity_field` computes the velocity components derived from the potential function using finite difference methods.

- `plot_flow` visualizes the potential field and superimposes streamline plots to denote flow patterns around objects.

The implementation assumes general boundary conditions, which can be modified for specific geometries, such as cylinders, by altering the `boundary_conditions` function.

Multiple Choice Questions

1. In potential flow theory, which equation describes the velocity field as a gradient of a scalar potential function?

    (a) Bernoulli's equation

    (b) Navier-Stokes equation

    (c) Laplace equation

    (d) Poisson equation

2. The Laplace equation $\nabla^2 \phi = 0$ is central to potential flow theory. What assumption primarily simplifies the fluid flow to meet this condition?

    (a) Viscous flow

    (b) Compressible flow

    (c) Irrotational flow

    (d) Transient flow

3. In the context of potential flow, what boundary condition is typically applied at solid surfaces?

    (a) Vanishing shear stress

    (b) No-slip condition

    (c) No-penetration condition

    (d) Constant pressure

4. For two-dimensional potential flows, the stream function ψ ensures which of the following?

    (a) Satisfaction of the momentum equation

    (b) Continuity equation is satisfied automatically

    (c) Flow is compressible

(d) Viscous effects are considered

5. In the complex potential $\Phi(z) = \phi + i\psi$, what does the derivative $\frac{d\Phi}{dz}$ represent?

    (a) Stream function
    (b) Pressure field
    (c) Complex velocity
    (d) Vorticity

6. Which potential flow solution describes the uniform flow around a cylindrical object?

    (a) Source-sink flow
    (b) Rankine vortex
    (c) Flow around a cylinder
    (d) Hele-Shaw flow

7. What is a primary limitation of potential flow theory in predicting real fluid behaviors?

    (a) Ignore energy conservation
    (b) Assumes fluid is compressible
    (c) Neglects viscous effects and flow separation
    (d) Assumes turbulent flow

Answers:

1. **C: Laplace equation** The velocity field in potential flow is described as the gradient of a scalar potential function that satisfies the Laplace equation $\nabla^2 \phi = 0$, a fundamental equation in potential flow theory.

2. **C: Irrotational flow** The assumption of irrotational flow simplifies the complex fluid motion equations and allows the derivation of the Laplace equation, central to potential flow theory.

3. **C: No-penetration condition** The no-penetration condition is applied at solid surfaces to ensure that the normal component of the velocity at the boundary is zero, maintaining the flow's tangential nature.

4. **B: Continuity equation is satisfied automatically** Defining the stream function in 2D flow setups ensures the continuity equation (representing mass conservation) is satisfied automatically, as the defined velocity components inherently honor this condition.

5. **C: Complex velocity** The derivative $\frac{d\Phi}{dz}$ gives the complex velocity, a powerful concept that combines the flow's potential and stream functions in two-dimensional analyses.

6. **C: Flow around a cylinder** The potential flow solution that models the flow around a cylinder involves the superposition of uniform flow and a doublet, resulting in characteristic patterns such as stagnation points on the cylinder surface.

7. **C: Neglects viscous effects and flow separation** Potential flow theory assumes inviscid, irrotational conditions, neglecting the effects of viscosity and flow separations, which are crucial in real-world fluid flow scenarios, especially near surfaces and in boundary layers.

Chapter 7

Pressure Poisson Equation

Derivation of the Pressure Poisson Equation

Within the domain of incompressible flow dynamics, the pressure Poisson equation is derived to address the constraint imposed by mass conservation. For incompressible flows, the continuity equation simplifies to:

$$\nabla \cdot \mathbf{u} = 0 \tag{7.1}$$

where $\mathbf{u}$ denotes the velocity vector field. The Navier-Stokes equations govern fluid motion and are expressed in vector form as:

$$\rho \left(\frac{\partial \mathbf{u}}{\partial t} + \mathbf{u} \cdot \nabla \mathbf{u} \right) = -\nabla p + \mu \nabla^2 \mathbf{u} + \mathbf{f} \tag{7.2}$$

where ρ is the fluid density, p is the pressure, μ is the dynamic viscosity, and $\mathbf{f}$ represents body forces.

To decouple pressure from the velocity field, the divergence of the momentum equation is taken:

$$\nabla \cdot \left(\frac{\partial \mathbf{u}}{\partial t} + \mathbf{u} \cdot \nabla \mathbf{u} \right) = -\nabla^2 p + \frac{\mu}{\rho} \nabla^2 (\nabla \cdot \mathbf{u}) + \nabla \cdot \mathbf{f} \tag{7.3}$$

Utilizing the incompressibility condition, the equation simplifies to the pressure Poisson formulation:

$$\nabla^2 p = \rho \nabla \cdot (\mathbf{u} \cdot \nabla \mathbf{u}) + \nabla \cdot \mathbf{f} \qquad (7.4)$$

This equation provides a framework to ensure pressure consistency for incompressible flows, facilitating the computation of pressure fields that satisfy both momentum and mass conservation laws.

Numerical Implementation in Fluid Simulations

The computational resolution of the pressure Poisson equation is integral in numerical solvers for incompressible flow problems, particularly in the context of the fractional step method. A typical computational framework involves discretizing the equation using techniques such as finite difference or finite volume methods.

Considerations include the application of appropriate boundary conditions and the deployment of iterative solvers, such as the Successive Over-Relaxation (SOR) or Multigrid methods, to achieve convergence effectively.

Boundary Conditions for Pressure Fields

Establishing boundary conditions for the pressure Poisson equation is critical to the accurate representation of physical scenarios. Common boundary treatments involve specifying pressure values or pressure gradients based on flow characteristics at the domain boundaries.

The no-penetration and no-slip conditions impacting velocity fields have indirect implications on the pressure distribution, which must be carefully incorporated in boundary formulations. This ensures the global consistency of the pressure solution, adhering to the physical constraints imposed by the domain's boundaries.

Role in Navier-Stokes Solvers

The pressure Poisson equation is pivotal in the projection method for incompressible Navier-Stokes solvers. The velocity field is first predicted without considering pressure effects, commonly referred to as the intermediate velocity field $\mathbf{u}^*$. This prediction does not inherently satisfy incompressibility.

By solving the pressure Poisson equation, the pressure field is determined, which is then used to correct the predicted velocity field. The corrected velocity field, $\mathbf{u}^{n+1}$, is computed as:

$$\mathbf{u}^{n+1} = \mathbf{u}^* - \frac{\Delta t}{\rho} \nabla p \qquad (7.5)$$

where Δt represents the time step. This correction ensures that the divergence-free condition is maintained at each time step, thereby preserving mass conservation.

Advanced Topics: Coupling with Turbulence Models

The integration of pressure Poisson solutions within turbulence modeling frameworks, such as Large Eddy Simulation (LES) or Reynolds-Averaged Navier-Stokes (RANS), enhances the fidelity of simulations. The interaction between pressure-driven forces and turbulent structures demands accurate resolution of pressure gradients and their influence on eddy formation and dissipation processes.

The pressure Poisson equation continues to be a foundational element in diverse computational fluid dynamics applications, spanning aerodynamics, hydrodynamics, and multiphase flow simulations.

Python Code Snippet

Below is a Python code snippet that encompasses the core computational elements for solving the Pressure Poisson equation, numerical implementations, and boundary condition considerations mentioned in this chapter.

```
import numpy as np

def pressure_poisson(p, dx, dy, rho, dt, u, v):
    '''
    Solve the pressure Poisson equation.
    :param p: Initial pressure field.
    :param dx: Grid spacing in x direction.
```

```
:param dy: Grid spacing in y direction.
:param rho: Fluid density.
:param dt: Time step.
:param u: Velocity in x direction.
:param v: Velocity in y direction.
:return: Updated pressure field.
'''
p_new = np.copy(p)
b = np.zeros_like(p)

# Compute the right-hand side of the Poisson equation
b[1:-1, 1:-1] = (rho * (1/dt *
                ((u[1:-1, 2:] - u[1:-1, 0:-2]) / (2*dx) +
                 (v[2:, 1:-1] - v[0:-2, 1:-1]) / (2*dy)) -
                ((u[2:, 1:-1] - u[0:-2, 1:-1]) / (2*dx))**2 -
                2*((u[1:-1, 2:] - u[1:-1, 0:-2]) / (2*dx) *
                   (v[2:, 1:-1] - v[0:-2, 1:-1]) / (2*dy)) -
                ((v[1:-1, 2:] - v[1:-1, 0:-2]) / (2*dy))**2))

# Iterate to solve for p
for _ in range(50):  # Assume convergence is reached in 50
↪   iterations
    p_new[1:-1, 1:-1] = (((dy**2 * (p_new[1:-1, 2:] +
    ↪   p_new[1:-1, 0:-2]) +
                          dx**2 * (p_new[2:, 1:-1] +
                          ↪   p_new[0:-2, 1:-1]) -
                          b[1:-1, 1:-1] * dx**2 * dy**2) /
                         (2 * (dx**2 + dy**2))))

    # Boundary conditions
    p_new[:, -1] = p_new[:, -2]    # dp/dx = 0 at x = 2
    p_new[0, :] = p_new[1, :]      # dp/dy = 0 at y = 0
    p_new[:, 0] = p_new[:, 1]      # dp/dx = 0 at x = 0
    p_new[-1, :] = 0               # p = 0 at y = 2

return p_new

def update_velocity(u, v, p, dx, dy, dt, rho, nu):
    '''
    Update velocity fields using pressure gradient.
    :param u: Velocity in x direction.
    :param v: Velocity in y direction.
    :param p: Pressure field.
    :param dx: Grid spacing in x direction.
    :param dy: Grid spacing in y direction.
    :param dt: Time step.
    :param rho: Fluid density.
    :param nu: Kinematic viscosity.
    :return: Updated velocity fields u and v.
    '''
    un = np.copy(u)
    vn = np.copy(v)
```

```python
        u[1:-1, 1:-1] = (un[1:-1, 1:-1] -
                        dt / dx * (p[1:-1, 2:] - p[1:-1, 0:-2]) / (2 *
                        ↪ rho) +
                        nu * (dt / dx**2 * (un[1:-1, 2:] - 2 * un[1:-1,
                        ↪ 1:-1] + un[1:-1, 0:-2]) +
                            dt / dy**2 * (un[2:, 1:-1] - 2 * un[1:-1,
                            ↪ 1:-1] + un[0:-2, 1:-1])))

        v[1:-1, 1:-1] = (vn[1:-1, 1:-1] -
                        dt / dy * (p[2:, 1:-1] - p[0:-2, 1:-1]) / (2 *
                        ↪ rho) +
                        nu * (dt / dx**2 * (vn[1:-1, 2:] - 2 * vn[1:-1,
                        ↪ 1:-1] + vn[1:-1, 0:-2]) +
                            dt / dy**2 * (vn[2:, 1:-1] - 2 * vn[1:-1,
                            ↪ 1:-1] + vn[0:-2, 1:-1])))

        # Boundary conditions for u and v
        u[0, :] = 0
        u[-1, :] = 0
        u[:, 0] = 0
        u[:, -1] = 0

        v[0, :] = 0
        v[-1, :] = 0
        v[:, 0] = 0
        v[:, -1] = 0

        return u, v

# Example parameters
nx, ny = 41, 41
dx = 2 / (nx - 1)
dy = 2 / (ny - 1)
p = np.zeros((ny, nx))          # Pressure field
u = np.zeros((ny, nx))          # Velocity in x direction
v = np.zeros((ny, nx))          # Velocity in y direction
rho = 1                         # Density
nu = 0.1                        # Viscosity
dt = 0.001                      # Time step

# Example of simulation step
p = pressure_poisson(p, dx, dy, rho, dt, u, v)
u, v = update_velocity(u, v, p, dx, dy, dt, rho, nu)

# Outputs for demonstration
print("Pressure field:\n", p)
print("Velocity field u:\n", u)
print("Velocity field v:\n", v)
```

This code provides a Python implementation for solving the Pressure Poisson equation and updating velocity fields in the context of incompressible flow simulations:

- `pressure_poisson` function solves the Pressure Poisson equation to compute the pressure field.
- `update_velocity` updates the velocity fields using pressure gradients and accounts for boundary conditions.
- Parameters such as grid spacing, time step, fluid properties, and initial conditions are defined for a simulation scenario.
- The simulation updates the pressure and velocity fields iteratively, representing a single simulation time step.

The final outputs demonstrate the computed pressure and velocity fields for a snapshot in the simulation, illustrating the role of the Pressure Poisson equation in maintaining incompressibility.

Multiple Choice Questions

1. The pressure Poisson equation is primarily used to:
   (a) Calculate fluid velocity directly
   (b) Ensure mass conservation in incompressible flows
   (c) Determine temperature distribution in fluids
   (d) Model compressible flow effects

2. In the derivation of the pressure Poisson equation, which condition is utilized?
   (a) Adiabatic condition
   (b) Isentropic condition
   (c) Incompressibility condition
   (d) Ideal gas condition

3. Which numerical method is commonly used to solve the pressure Poisson equation in fluid dynamics simulations?
   (a) Finite Element Method (FEM)
   (b) Finite Difference Method (FDM)
   (c) Monte Carlo Method
   (d) Lagrangian Method

4. What role does the pressure Poisson equation play in the projection method for solving Navier-Stokes equations?

    (a) It predicts the temperature field
    (b) It calculates viscosity effects
    (c) It corrects the velocity field to ensure divergence-free condition
    (d) It models boundary interactions

5. Boundary conditions for the pressure Poisson equation are essential because they:

    (a) Simplify the computational grid
    (b) Allow direct calculation of turbulence intensity
    (c) Ensure accurate representation of physical scenarios and pressure distributions
    (d) Reduce computational cost

6. An important aspect when integrating the pressure Poisson equation in turbulence models is:

    (a) Reducing computational time
    (b) Enhancing the simulation's stability
    (c) Accurately resolving pressure gradients to influence eddy dynamics
    (d) Simplifying the mesh complexity

7. In numerical simulations, the iterative solvers used for the pressure Poisson equation, like Multigrid, are chosen mainly due to their:

    (a) Ability to handle non-linear equations directly
    (b) Efficiency and speed in solving large, sparse linear systems
    (c) Compatibility with commercial software
    (d) High precision in predicting vortex formation

Answers:
1. **B: Ensure mass conservation in incompressible flows**
The pressure Poisson equation is specifically employed to ensure

that the incompressibility condition, or mass conservation, is satisfied in fluid simulations.

2. **C: Incompressibility condition** The derivation of the pressure Poisson equation makes use of the incompressibility condition, $\nabla \cdot \mathbf{u} = 0$, to decouple pressure from the velocity field.

3. **B: Finite Difference Method (FDM)** The Finite Difference Method is one of the common numerical methods used to discretize and solve the pressure Poisson equation in fluid dynamics problems.

4. **C: It corrects the velocity field to ensure divergence-free condition** In the projection method, the pressure Poisson equation is solved to correct the predicted velocity so that it satisfies the divergence-free condition for incompressible flows.

5. **C: Ensure accurate representation of physical scenarios and pressure distributions** Proper boundary conditions for the pressure Poisson equation ensure that the simulated pressure field accurately reflects the physical boundaries and conditions of the system.

6. **C: Accurately resolving pressure gradients to influence eddy dynamics** In turbulence models, the correct resolution of pressure gradients is crucial for accurately capturing the effects of turbulent eddies in the fluid flow.

7. **B: Efficiency and speed in solving large, sparse linear systems** Iterative solvers like the Multigrid method are favored for their efficiency and speed when solving the large, sparse linear systems formed by the discretization of the pressure Poisson equation.

Chapter 8

Energy Equations for Compressible Fluids

Governing Equations

In the arena of compressible fluid dynamics, the energy equation assumes a pivotal role in capturing the interplay between heat transfer and work interactions. Starting with the fundamental principles, the energy equation for compressible flows is derived from the first law of thermodynamics, encapsulated within the conservative form:

$$\frac{\partial}{\partial t}(\rho e_T) + \nabla \cdot (\rho e_T \mathbf{u}) = -\nabla \cdot \mathbf{q} + \rho \dot{q}'' + \nabla \cdot (\tau \cdot \mathbf{u}) + \rho \dot{w} \quad (8.1)$$

where ρ is the fluid density, $e_T = e + \frac{1}{2}u^2$ is the total energy per unit mass including internal energy e and kinetic energy, $\mathbf{q}$ denotes the heat flux vector, $\dot{q}''$ embodies volumetric heat addition (or removal), τ represents the stress tensor, and $\dot{w}$ is the specific work done by body forces.

Internal Energy Models

The internal energy e is a function of specific heat capacities at constant volume, speculative in relation to temperature and specific volume alterations in compressible systems. The ideal gas approximation provides a simplified model:

$$e = c_v T \qquad (8.2)$$

where c_v stands for the specific heat at constant volume, often necessitating corrections for non-ideal behaviors at elevated pressures and temperatures.

Heat Transfer Modeling

The Fourier law governs the conductive heat transfer, integrating seamlessly into the compressible flow framework:

$$\mathbf{q} = -k\nabla T \qquad (8.3)$$

where k is the thermal conductivity of the fluid. Within convective regimes, dimensionless parameters such as the Nusselt number are instrumental in linking heat transfer rates to flow dynamics.

Work Interactions

Work interactions within compressible flows demand careful consideration of viscous effects and flow-induced stresses. The mechanical energy equation isolates the work done by pressure forces and viscous stresses:

$$\frac{\partial}{\partial t}\left(\frac{1}{2}\rho u^2\right) + \nabla \cdot \left(\frac{1}{2}\rho u^2 \mathbf{u}\right) = -\mathbf{u} \cdot \nabla p + \nabla \cdot (\tau \cdot \mathbf{u}) \qquad (8.4)$$

The dual role of pressure gradients exacerbates compressibility effects, altering both acceleration fields and thermodynamic states.

Entropy Considerations

The second law of thermodynamics imposes an additional constraint, expressed through the entropy transport equation:

$$\frac{\partial}{\partial t}(\rho s) + \nabla \cdot (\rho s \mathbf{u}) = \frac{\rho \dot{q}''}{T} + \Phi \qquad (8.5)$$

with s as the specific entropy and Φ denoting the dissipation function, capturing irreversible processes from viscosity-induced internal friction.

Numerical Approaches

Implementing energy equations numerically relies heavily on discretization techniques, such as the finite volume method (FVM), and advanced solver methods to address nonlinearities inherent in compressible flows. Iterative solvers and advanced algorithms, such as the preconditioned conjugate gradient technique, enhance computational stability and efficiency.

The balance of resolving heat transfer and mechanical work in tandem demands high fidelity in mesh construction and temporal resolution, ensuring convergence and accuracy.

Python Code Snippet

Below is a Python code snippet that encompasses the core computational elements for solving the governing energy equations for compressible fluids, including heat transfer and work interactions.

```python
import numpy as np

def total_energy_density(rho, e, u):
    '''
    Calculate the total energy density.
    :param rho: Fluid density.
    :param e: Internal energy per unit mass.
    :param u: Velocity vector (numpy array).
    :return: Total energy density.
    '''
    return rho * (e + 0.5 * np.dot(u, u))

def heat_flux(k, grad_T):
    '''
    Compute the heat flux vector using Fourier's law.
    :param k: Thermal conductivity.
    :param grad_T: Temperature gradient vector (numpy array).
    :return: Heat flux vector.
    '''
    return -k * grad_T

def work_done_by_body_forces(rho, u, tau, body_force_work):
    '''
    Calculate the work done by body forces in compressible flow.
    :param rho: Fluid density.
    :param u: Velocity vector (numpy array).
    :param tau: Stress tensor.
    :param body_force_work: Specific work done by body forces.
    :return: Total work term.
```

```python
    '''
    return np.dot(tau, u) + rho * body_force_work

def update_entropy(rho, s, specific_heat, temperature,
    heat_addition, dissipation_function):
    '''
    Update the entropy of the system based on second law of
        thermodynamics.
    :param rho: Fluid density.
    :param s: Specific entropy.
    :param specific_heat: Specific heat capacity.
    :param temperature: Temperature.
    :param heat_addition: Volumetric heat addition.
    :param dissipation_function: Dissipation function.
    :return: Updated entropy.
    '''
    return s + (rho * heat_addition / temperature) +
        dissipation_function

def entropy_transport_equation(rho, u, s, heat_addition,
    temperature, dissipation_function):
    '''
    Solve the entropy transport equation.
    :param rho: Fluid density.
    :param u: Velocity vector (numpy array).
    :param s: Specific entropy.
    :param heat_addition: Volumetric heat addition.
    :param temperature: Temperature.
    :param dissipation_function: Dissipation function.
    :return: Change in entropy.
    '''
    return rho * np.dot(u, s) + (rho * heat_addition / temperature)
        + dissipation_function

def numerically_solve_energy_equation(parameters):
    '''
    Simulate the solver for the compressible energy equation under
        specified conditions.
    :param parameters: Dictionary containing parameters such as
        density, velocity, heat flux, etc.
    :return: Results of the numerical solution.
    '''
    # This function is a placeholder for evolving the energy
    #     equation over time using discrete methods
    # Initialize variables from `parameters`
    rho = parameters['rho']
    e = parameters['e']
    u = parameters['u']
    tau = parameters['tau']
    body_force_work = parameters['body_force_work']
    grad_T = parameters['grad_T']
    k = parameters['k']
    s = parameters['s']
```

```python
    temperature = parameters['temperature']
    heat_addition = parameters['heat_addition']
    dissipation_function = parameters['dissipation_function']

    # Calculate various terms using functions
    e_density = total_energy_density(rho, e, u)
    q = heat_flux(k, grad_T)
    work = work_done_by_body_forces(rho, u, tau, body_force_work)
    entropy_change = entropy_transport_equation(rho, u, s,
    ↪   heat_addition, temperature, dissipation_function)

    # Return computed results
    return e_density, q, work, entropy_change

# Parameters initialization
parameters = {
    'rho': 1.0,
    'e': 5.0,
    'u': np.array([2.0, 0.0, 0.0]),
    'tau': np.array([0.0, 0.0, 0.0]),
    'body_force_work': 0.0,
    'grad_T': np.array([-0.5, 0.0, 0.0]),
    'k': 0.02,
    's': 0.9,
    'temperature': 300.0,
    'heat_addition': 0.1,
    'dissipation_function': 0.05
}

# Simulation call
energy_density, heat_flux_vector, work_term, entropy_change =
↪   numerically_solve_energy_equation(parameters)

print("Energy Density:", energy_density)
print("Heat Flux Vector:", heat_flux_vector)
print("Work Done:", work_term)
print("Entropy Change:", entropy_change)
```

This code defines several key functions to evaluate energy dynamics in compressible fluid flow systems:

- `total_energy_density` computes the total energy density of the fluid using its density and velocity.

- `heat_flux` utilizes Fourier's law to calculate conductive heat transfer.

- `work_done_by_body_forces` accounts for the work done by various forces in the flow, including stress-induced actions.

- `update_entropy` and `entropy_transport_equation` model entropy changes based on the second law of thermodynamics.

- `numerically_solve_energy_equation` is a placeholder for solving the compressible energy equation using specified parameters for simulation purposes.

This comprehensive block of code demonstrates the computational aspects critical for analyzing thermodynamics in compressible fluid dynamics.

Multiple Choice Questions

1. Which term describes the total energy per unit mass in the energy equation for compressible flows?

    (a) Thermal energy

    (b) Kinetic energy

    (c) Total energy including potential energy

    (d) Total energy including internal energy

2. What role does the Fourier law play in compressible fluid dynamics?

    (a) It governs the conductive heat transfer.

    (b) It models the convective heat transfer.

    (c) It describes electromagnetic interactions.

    (d) It explains the relation of viscosity with velocity gradients.

3. Which of the following is not directly part of the energy equation for compressible fluid flows?

    (a) Heat flux vector

    (b) Stress tensor

    (c) Electric potential

    (d) Body force work

4. How is the specific heat at constant volume denoted in the internal energy expression for an ideal gas?

    (a) c_p

(b) c_v

(c) γ

(d) R

5. What does the entropy transport equation primarily account for in fluid dynamics?

    (a) Mass conservation

    (b) Heat addition

    (c) Entropy consistency and dissipation

    (d) Pressure equilibrium

6. When using numerical approaches for energy equations, what is a typical method for discretization?

    (a) Finite difference method

    (b) Finite volume method

    (c) Euler's method

    (d) Trapezoidal rule

7. What is a primary challenge in solving energy equations for compressible flows numerically?

    (a) Handling low-speed regimes

    (b) Ensuring mass conservation

    (c) Addressing nonlinearities and stability

    (d) Managing incompressibility

Answers:

1. **D: Total energy including internal energy** The total energy per unit mass in the energy equation includes both internal and kinetic energy components, critical for compressible fluid flow analysis.

2. **A: It governs the conductive heat transfer.** The Fourier law is employed for modeling conductive heat transfer, describing how temperature gradients drive heat flow in a material.

3. **C: Electric potential** The energy equation for compressible flows does not directly involve electric potential, focusing instead on heat, work interactions, and fluid dynamics.

4. **B:** c_v In the internal energy expression for an ideal gas, c_v represents the specific heat at constant volume.

5. **C: Entropy consistency and dissipation** The entropy transport equation accounts for entropy changes within a system, including generating due to irreversible processes like friction.

6. **B: Finite volume method** The finite volume method is commonly used for discretizing the equations governing fluid flows, particularly suited for conservation laws in computational fluid dynamics.

7. **C: Addressing nonlinearities and stability** Numerical solutions to compressible flow energy equations are complicated by inherent nonlinearities and stability considerations, necessitating sophisticated solver approaches.

Chapter 9

The Euler Equations

Introduction to the Euler Equations

The Euler equations serve as the cornerstone for analyzing inviscid flows, a critical simplification under the broader domain of fluid mechanics. In the absence of viscous effects, these equations govern the motion of fluid elements by simulating an idealized environment where shear stresses and heat conduction are negligible. The Euler equations can be succinctly expressed within a conservative form:

$$\frac{\partial \rho}{\partial t} + \nabla \cdot (\rho \mathbf{u}) = 0, \tag{9.1}$$

$$\frac{\partial (\rho \mathbf{u})}{\partial t} + \nabla \cdot (\rho \mathbf{u} \otimes \mathbf{u} + p\mathbf{I}) = 0, \tag{9.2}$$

$$\frac{\partial E}{\partial t} + \nabla \cdot ((E + p)\mathbf{u}) = 0, \tag{9.3}$$

where ρ denotes the fluid density, $\mathbf{u}$ represents the velocity vector, p indicates the pressure, $\mathbf{I}$ is the identity matrix, and E refers to the total energy per unit volume. These equations encapsulate mass, momentum, and energy conservation laws respectively.

Assumptions and Simplifications

Distilling the dynamics of inviscid flows requires foundational assumptions which decouple these analyses from the complications of viscosity. The Euler equations inherently stipulate an ideal gas assumption, where the flow regime is uninfluenced by frictional forces:

- Negligible viscosity ($\mu \approx 0$)
- No thermal conductivity ($k_{thermal} \approx 0$)
- Pressure derivation from ideal gas law: $p = \rho RT$

This framework is particularly viable in high-speed aerodynamics, where viscous effects are overshadowed by inertial forces.

Applications in Aerospace Engineering

Deploying the Euler equations within aerospace engineering underscores their pivotal role in shaping aircraft design and performance assessments. Computational models leverage these equations to simulate airflows around airfoils and fuselage, evaluating lift and drag coefficients. The pressure distribution, gleaned from the Euler solution, influences the aerodynamic qualities of an aircraft.

The critical Mach number dictating compressibility effects is derived from:

$$M = \frac{u}{a} = \frac{u}{\sqrt{\gamma RT}}, \qquad (9.4)$$

where M is the Mach number, u the flow velocity, a the speed of sound, and γ the specific heat ratio. This evaluation directs design considerations vis-à-vis shock waves and expansion fans for supersonic crafts.

Relevance to Aerodynamic Design

The precision in developing aerodynamic models hinges significantly on the Euler framework, particularly in scenarios demanding rapid prototyping and testing. Simplified inviscid analyses afford engineers a preliminary investigation into flow behaviour over construct surfaces, enabling rapid iterations in the design phase. Key aerodynamic parameters such as:

$$C_p = \frac{p - p_\infty}{0.5 \rho_\infty u_\infty^2}, \qquad (9.5)$$

$$C_d = \frac{D}{0.5 \rho_\infty u_\infty^2 A}, \qquad (9.6)$$

where C_p represents the pressure coefficient, C_d the drag coefficient, D the drag force, ρ_∞ the freestream density, and u_∞ the freestream velocity, are appraised using solutions to the Euler equations, enhancing optimization strategies.

Numerical Simulation Techniques

Numerical simulations augment theoretical explorations, embodying a core aspect of studying inviscid flows. Finite difference, finite volume, and finite element methods manifest as the primary computational schemes employed in solving Euler equations. These numerical techniques necessitate high-fidelity spatial discretization to capture shock and expansion phenomena accurately, crucial for fidelity in supersonic and hypersonic flow assessments.

Incorporating time-stepping algorithms such as Runge-Kutta schemes or predictor-corrector methods propels the dynamic modeling capabilities, accommodating transient flow phenomena across diverse aerospace applications. The evolution of computational fluid dynamics (CFD) must continually harness optimized solvers and enhanced mesh techniques to resolve the nuances entailed by inviscid analyses effectively.

Python Code Snippet

Below is a Python code snippet that encompasses the core computational elements related to the Euler equations, providing functions for calculating the Mach number, pressure and drag coefficients, and simulating basic inviscid flow dynamics using numerical techniques.

```
import numpy as np

def calculate_mach_number(u, R, T, gamma):
    '''
    Calculate the Mach number.
    :param u: Flow velocity.
    :param R: Gas constant.
    :param T: Temperature.
    :param gamma: Specific heat ratio.
    :return: Mach number.
    '''
    a = np.sqrt(gamma * R * T)
    return u / a
```

```python
def pressure_coefficient(p, p_inf, rho_inf, u_inf):
    '''
    Calculate the pressure coefficient.
    :param p: Pressure at a point.
    :param p_inf: Freestream pressure.
    :param rho_inf: Freestream density.
    :param u_inf: Freestream velocity.
    :return: Pressure coefficient.
    '''
    return (p - p_inf) / (0.5 * rho_inf * u_inf**2)

def drag_coefficient(D, rho_inf, u_inf, A):
    '''
    Calculate the drag coefficient.
    :param D: Drag force.
    :param rho_inf: Freestream density.
    :param u_inf: Freestream velocity.
    :param A: Reference area.
    :return: Drag coefficient.
    '''
    return D / (0.5 * rho_inf * u_inf**2 * A)

# Example parameters for a simulated inviscid flow
params = {
    'u': 300, 'R': 287, 'T': 288, 'gamma': 1.4, 'p': 101325,
    'p_inf': 100000, 'rho_inf': 1.225, 'u_inf': 340, 'D': 500, 'A':
    ↪   2
}

# Numerical simulation with basic time-stepping algorithms
def simulate_inviscid_flow(params, time_steps=100, dt=0.01):
    '''
    Simulate inviscid flow using Euler equations with a simple
    ↪   numerical solver.
    :param params: Dictionary of flow parameters.
    :param time_steps: Number of time steps for the simulation.
    :param dt: Time step duration.
    :return: Array of calculated Mach numbers over time.
    '''
    mach_numbers = np.zeros(time_steps)
    for t in range(time_steps):
        mach_numbers[t] = calculate_mach_number(
            params['u'], params['R'], params['T'], params['gamma']
        )
        # Simulate updating flow conditions (simplified)
        params['u'] -= 0.01 * t  # Example: Decreasing velocity over
        ↪   time

    return mach_numbers

# Outputs for demonstration
```

```
mach_num = calculate_mach_number(params['u'], params['R'],
↪   params['T'], params['gamma'])
C_p = pressure_coefficient(params['p'], params['p_inf'],
↪   params['rho_inf'], params['u_inf'])
C_d = drag_coefficient(params['D'], params['rho_inf'],
↪   params['u_inf'], params['A'])
mach_numbers_over_time = simulate_inviscid_flow(params)

print("Mach Number:", mach_num)
print("Pressure Coefficient:", C_p)
print("Drag Coefficient:", C_d)
print("Mach Numbers Over Time:", mach_numbers_over_time)
```

This code defines several key functions essential for evaluating parameters critical in fluid dynamics and aerodynamics:

- `calculate_mach_number` computes the Mach number based on flow speed, gas constant, temperature, and specific heat ratio.

- `pressure_coefficient` calculates the pressure coefficient, which is essential for understanding flow behavior around objects.

- `drag_coefficient` provides the drag coefficient based on flow conditions and reference area.

- `simulate_inviscid_flow` demonstrates a simple numerical method to simulate inviscid flows, iterating through time steps to show varying Mach number dynamics.

The final block of code provides examples of applying these functions using sample data, illustrating how Mach number and coefficients change in an inviscid environment.

Multiple Choice Questions

1. Which of the following assumptions is NOT inherent to the Euler equations when applied to fluid dynamics?

    (a) Negligible viscosity

    (b) No thermal conductivity

    (c) Constant density

    (d) Ideal gas behavior

2. The Euler equations are primarily used in the analysis of which type of fluid flows?

    (a) Compressible and viscous
    (b) Inviscid and compressible
    (c) Viscous and incompressible
    (d) Inviscid and incompressible

3. The expression $\frac{\partial \rho}{\partial t} + \nabla \cdot (\rho \mathbf{u}) = 0$ corresponds to which conservation law?

    (a) Conservation of momentum
    (b) Conservation of energy
    (c) Conservation of mass
    (d) Conservation of vorticity

4. In the context of the Euler equations, the Mach number is crucial for evaluating:

    (a) Viscous effects
    (b) Shear stresses
    (c) Compressibility effects
    (d) Thermal conduction

5. Which numerical method is frequently used to solve the Euler equations for high-speed aerodynamics?

    (a) Finite difference method
    (b) Boundary element method
    (c) Direct numerical simulation
    (d) Arbitrary Lagrangian-Eulerian method

6. What is the primary effect of implementing the Euler equations in aerodynamic design?

    (a) Increased computational complexity
    (b) Assessment of lift and drag coefficients
    (c) Detailed analysis of temperature gradients
    (d) Comprehensive evaluation of viscous effects

7. The equation $C_p = \frac{p-p_\infty}{0.5\rho_\infty u_\infty^2}$ is used to determine which aerodynamic parameter?

    (a) Drag coefficient
    (b) Pressure coefficient
    (c) Lift coefficient
    (d) Friction coefficient

Answers:

1. **C: Constant density** The assumption of constant density is not inherent in the Euler equations. They address inviscid flows where density can vary, especially in compressible regimes.

2. **B: Inviscid and compressible** The Euler equations cater to inviscid flows, which can be both compressible and incompressible, but are particularly pivotal in scenarios with significant compressibility effects, like high-speed flows.

3. **C: Conservation of mass** This equation is the continuity equation, representing the conservation of mass principle in fluid dynamics.

4. **C: Compressibility effects** The Mach number is a critical parameter in determining when compressibility effects in a fluid flow are significant.

5. **A: Finite difference method** The finite difference method, along with finite volume and finite element methods, is commonly used to discretize the Euler equations in computational fluid dynamics.

6. **B: Assessment of lift and drag coefficients** The Euler equations are utilized in evaluating lift and drag forces, which are key considerations in the aerodynamic design of aircraft.

7. **B: Pressure coefficient** This equation defines the pressure coefficient, a dimensionless number that characterizes the relative pressure at a point in a fluid flow.

Chapter 10

Boundary Layer Equations

Introduction to Boundary Layer Theory

The concept of the boundary layer is a fundamental aspect of fluid mechanics, introduced by Ludwig Prandtl in the early 20th century. It describes the thin region adjacent to a solid surface where viscous effects are significant in an otherwise inviscid flow. The governing equations of the boundary layer are derived by simplifying the Navier-Stokes equations under the assumption of a high Reynolds number, where inertial forces overwhelmingly dominate over viscous forces outside the boundary layer.

Formulation of Laminar Boundary Layer Equations

In the laminar regime, the boundary layer equations arise from a balance where viscous diffusion and convection are the primary contributors. The Prandtl's boundary layer equations in two dimensions for steady incompressible flow are expressed as follows:

$$\frac{\partial u}{\partial x} + \frac{\partial v}{\partial y} = 0, \tag{10.1}$$

$$u\frac{\partial u}{\partial x} + v\frac{\partial u}{\partial y} = -\frac{1}{\rho}\frac{\partial p}{\partial x} + \nu\frac{\partial^2 u}{\partial y^2}, \tag{10.2}$$

where u and v are the velocity components in the x and y directions, respectively, ρ is the fluid density, p is pressure, and ν is the kinematic viscosity.

Turbulent Boundary Layer Formulation

For turbulent boundary layers, complexities arise due to the chaotic motion of fluid particles. The Reynolds-averaged Navier-Stokes (RANS) equations become applicable. The time-averaged form of the boundary layer equations are given by:

$$\frac{\partial \bar{u}}{\partial x} + \frac{\partial \bar{v}}{\partial y} = 0, \tag{10.3}$$

$$\bar{u}\frac{\partial \bar{u}}{\partial x} + \bar{v}\frac{\partial \bar{u}}{\partial y} = -\frac{1}{\rho}\frac{\partial \bar{p}}{\partial x} + \frac{\partial}{\partial y}\left(\nu\frac{\partial \bar{u}}{\partial y} - \overline{u'v'}\right), \tag{10.4}$$

where $\bar{u}$ and $\bar{v}$ denote time-averaged velocity components, and the term $\overline{u'v'}$ represents the Reynolds stress, embodying the momentum transfer due to turbulence.

Similarity Solutions for Laminar Boundary Layers

Similarity solutions offer precise solutions under specific flow conditions. The Blasius solution is a notable form applicable to flow over a flat plate. The dimensionless form of the governing equation is given by:

$$f''' + \frac{1}{2}ff'' = 0, \tag{10.5}$$

where f is the non-dimensional stream function, subject to the boundary conditions $f(0) = f'(0) = 0$ and $f'(\infty) = 1$.

Momentum Integral Equation in Boundary Layers

The momentum integral method provides an approximate solution to boundary layer problems, encapsulating the essence of the boundary layer growth. The Von Karman momentum integral equation for a steady, incompressible flow is represented as:

$$\frac{d}{dx}(\delta^* U) = \theta \frac{dU}{dx} + \tau_w, \tag{10.6}$$

where δ^* is the displacement thickness, θ is the momentum thickness, U is the outer flow velocity, and τ_w is the wall shear stress.

Analysis and Estimation of Turbulent Boundary Layers

Estimating turbulent boundary layer parameters necessitates empirical correlations due to the stochastic nature of turbulence. Skin friction coefficient C_f and boundary layer thickness δ predictions are often modeled using:

$$C_f = 0.0592 \left(\frac{Re_x}{L}\right)^{-0.2}, \tag{10.7}$$

$$\delta \approx 0.37x \left(\frac{Re_x}{L}\right)^{-0.2}, \tag{10.8}$$

where Re_x denotes the Reynolds number based on the distance from the leading edge x and characteristic length L.

Computational Techniques for Boundary Layer Analysis

Computational methods for analyzing boundary layers, such as finite difference and finite volume methods, incorporate numerical stability and convergence properties. Discretizing the boundary layer equations involves transforming the continuous equations into a solvable algebraic form. Typical approaches approximate derivatives using schemes like:

$$\left(\frac{\partial u}{\partial x}\right)_i \approx \frac{u_{i+1} - u_i}{\Delta x}, \tag{10.9}$$

$$\left(\frac{\partial^2 u}{\partial y^2}\right)_j \approx \frac{u_{j+1} - 2u_j + u_{j-1}}{\Delta y^2}, \tag{10.10}$$

where Δx and Δy are discretized grid spacings in the x and y directions.

Python Code Snippet

Below is a Python code snippet that encompasses the core computational elements of boundary layer analysis, including laminar and turbulent formulations, similarity solutions, and momentum integral method.

```python
import numpy as np
from scipy.integrate import solve_ivp

def laminar_boundary_layer(u_inf, nu, delta_y, x_range, y_range):
    '''
    Solves laminar boundary layer equations using finite difference
    ↪   method.
    :param u_inf: Freestream velocity.
    :param nu: Kinematic viscosity.
    :param delta_y: Discretization in y direction.
    :param x_range: Range of x domain.
    :param y_range: Range of y domain.
    :return: Velocity field u over the domain.
    '''
    # Initialize velocity field
    u = np.zeros((len(x_range), len(y_range)))

    # Boundary condition at y=0 (no-slip condition)
    u[:, 0] = 0

    # Freestream boundary condition
    u[:, -1] = u_inf

    # Marching in x direction
    for i in range(1, len(x_range)):
        for j in range(1, len(y_range) - 1):
            u[i, j] = u[i-1, j] + nu * delta_y * (u[i-1, j+1] -
            ↪   2*u[i-1, j] + u[i-1, j-1])

    return u

def turbulent_boundary_layer(u_inf, rho, nu, x_range, y_range):
    '''
    Solves turbulent boundary layer using empirical correlations.
    :param u_inf: Freestream velocity.
    :param rho: Fluid density.
    :param nu: Kinematic viscosity.
    :param x_range: Range of x domain.
    :param y_range: Range of y domain.
```

```
:return: Skin friction coefficient and boundary layer thickness.
'''
Re_x = (rho * u_inf * x_range) / nu
cf = 0.0592 * Re_x**-0.2
delta = 0.37 * x_range * Re_x**-0.2

return cf, delta

def blasius_solution(x_range):
    '''
    Computes the Blasius solution for laminar flow over a flat
    ↪ plate.
    :param x_range: Range of x domain.
    :return: Blasius velocity profile.
    '''
    def blasius_eq(eta, f):
        return [f[1], f[2], -0.5 * f[0] * f[2]]

    blasius_bc = [0, 0, 1]  # Boundary conditions at eta = 0

    # Solve Blasius equation using initial value solver
    solution = solve_ivp(blasius_eq, [0, max(x_range)], blasius_bc,
    ↪     dense_output=True)

    return solution.sol(x_range)[0]

# Example usage of functions:
u_inf = 1.0  # Freestream velocity
nu = 1.5e-5  # Kinematic viscosity
x_range = np.linspace(0, 5, 100)
y_range = np.linspace(0, 0.05, 50)

u_laminar = laminar_boundary_layer(u_inf, nu, y_range[1] -
↪     y_range[0], x_range, y_range)
cf, delta = turbulent_boundary_layer(u_inf, 1.0, nu, x_range,
↪     y_range)
f_blasius = blasius_solution(np.linspace(0, 5, 100))

print("Laminar Boundary Layer Velocity Field:", u_laminar)
print("Turbulent Skin Friction Coefficient:", cf)
print("Turbulent Boundary Layer Thickness:", delta)
print("Blasius Velocity Profile:", f_blasius)
```

This code defines several key functions necessary for the implementation of boundary layer analysis:

- `laminar_boundary_layer` function uses a finite difference method to solve laminar boundary layer equations.

- `turbulent_boundary_layer` applies empirical correlations to determine skin friction coefficient and boundary layer thickness in turbulent flows.

- `blasius_solution` computes the Blasius solution for a two-dimensional laminar boundary layer over a flat plate.

The final block of code provides example calculations using typical fluid flow parameters.

Multiple Choice Questions

1. Who initially introduced the concept of the boundary layer in fluid mechanics?

    (a) Hermann von Helmholtz

    (b) Richard Feynman

    (c) Bernoulli

    (d) Ludwig Prandtl

2. What assumption allows the simplification of the Navier-Stokes equations in the boundary layer theory?

    (a) Low Reynolds number

    (b) High Reynolds number

    (c) Linear flow regime

    (d) Non-Newtonian fluid behavior

3. In laminar boundary layer formulation, which forces primarily balance each other?

    (a) Inertial and gravitational forces

    (b) Viscous diffusion and convection

    (c) Centripetal and centrifugal forces

    (d) Coriolis and Lorentz forces

4. In the context of the turbulent boundary layer, what does the term $\overline{u'v'}$ represent?

    (a) Vorticity

    (b) Velocity fluctuation

    (c) Shear stress

    (d) Reynolds stress

5. What is the Blasius solution specifically applicable to?

(a) Flow around a cylinder

(b) Flow over a flat plate

(c) Flow through a nozzle

(d) Flow in a pipe

6. Which mathematical method provides an approximate solution for boundary layer growth?

(a) Blasius integration

(b) Runge-Kutta method

(c) Momentum integral method

(d) Finite volume method

7. Which empirical correlation is used to estimate skin friction coefficient C_f for a turbulent boundary layer?

(a) $C_f = 0.0791 Re_x^{-0.25}$

(b) $C_f = 0.0592 Re_x^{-0.2}$

(c) $C_f = 0.066 Re_x^{-0.3}$

(d) $C_f = 0.0471 Re_x^{-0.5}$

Answers:

1. **D: Ludwig Prandtl** Ludwig Prandtl is credited with introducing the boundary layer concept, which describes the thin layer of fluid in contact with a solid surface where viscous effects are significant.

2. **B: High Reynolds number** The assumption of a high Reynolds number allows the simplification of the Navier-Stokes equations in boundary layer theory since inertial forces dominate over viscous forces outside the boundary layer.

3. **B: Viscous diffusion and convection** In laminar boundary layers, viscous diffusion and convection are the main balancing forces, leading to the development of classical boundary layer equations.

4. **D: Reynolds stress** The Reynolds stress $\overline{u'v'}$ represents the additional stress due to turbulence fluctuations in the velocity field, crucial for modeling turbulent boundary layers.

5. **B: Flow over a flat plate** The Blasius solution provides a similarity solution for the laminar boundary layer developing over a flat plate in steady flow.

6. **C: Momentum integral method** The momentum integral method offers an approximate analysis for the growth of boundary layers, especially where exact solutions are difficult to obtain.

7. **B:** $C_f = 0.0592 Re_x^{-0.2}$ This empirical correlation is commonly used to predict the skin friction coefficient in turbulent boundary layers based on the Reynolds number.

Chapter 11

Conservation of Vorticity

Introduction to Vorticity Dynamics

Vorticity is a fundamental concept in fluid mechanics, providing insight into the rotational behavior of fluid elements. Defined mathematically as the curl of the velocity field, vorticity analyzes the local spinning motion of the fluid. The vorticity vector $\vec{\omega}$ is expressed as:

$$\vec{\omega} = \nabla \times \vec{v}$$

where $\vec{v}$ is the velocity vector of the fluid. The vorticity conservation principle arises from the Navier-Stokes equations by examining the transport of the vorticity field in a fluid domain.

Vorticity Equation Derivation

In an incompressible viscous fluid, the vorticity transport equation is derived from the Navier-Stokes equations and is given by:

$$\frac{\partial \vec{\omega}}{\partial t} + \vec{v} \cdot \nabla \vec{\omega} = \vec{\omega} \cdot \nabla \vec{v} + \nu \nabla^2 \vec{\omega}$$

where ν denotes the kinematic viscosity, and $\vec{v} \cdot \nabla \vec{\omega}$ represents the convection of vorticity. The term $\vec{\omega} \cdot \nabla \vec{v}$ signifies the vortex stretching and tilting, while $\nu \nabla^2 \vec{\omega}$ indicates viscous diffusion of vorticity.

Implications of Vorticity Conservation

Vorticity conservation implies that in inviscid flows, where $\nu = 0$, the vorticity of a fluid element is conserved along its pathline, subject to initial conditions. This aspect is crucial in understanding vortex dynamics, such as the persistence and evolution of vortices in a flow field and the concept of circulation Γ, defined by:

$$\Gamma = \oint_C \vec{v} \cdot d\vec{s}$$

where C is a closed contour in the flow. According to the Helmholtz theorems, the circulation remains constant for an inviscid and barotropic fluid in the absence of body forces.

Vortex Stretching and Tilting

The term $\vec{\omega} \cdot \nabla \vec{v}$ elaborates on the vortex interaction with the velocity field, leading to stretching and tilting phenomena. Vortex stretching occurs when a fluid element is elongated in the direction of the vorticity vector, resulting in an increase in the magnitude of vorticity. Conversely, vortex tilting illustrates the inclination modification of the vorticity vector due to the velocity field's gradient. These dynamics are crucial in turbulence and spiral formation in various fluid contexts.

Role in Turbulent Flows

In turbulent flows, vorticity dynamics play a significant role in energy transfer across scales. Vorticity augments the complexity of turbulent flow structures, where small-scale eddies are characterized by intense vorticity concentration. The `Reynolds-Averaged Navier-Stokes (RANS)` methods for turbulence modeling incorporate vorticity to comprehend the detailed interaction between turbulent eddies and mean flow structures.

Applications in Engineering

Understanding vorticity conservation principles aids in analyzing and designing various engineering systems. For instance, in aerodynamic applications, the control of vorticity is pivotal in minimizing drag and enhancing lift on airfoils. Vortex generators and aerodynamic surfaces are engineered to influence vorticity distribution, thereby refining the aircraft's performance parameters.

In conclusion, the conservation of vorticity serves as a cornerstone in comprehending complex fluid dynamics phenomena. Its implications span theoretical frameworks and practical applications, enabling the progression of both fundamental and applied fluid mechanics. Detailed vorticity analysis continues to illuminate intricate mechanisms governing macroscale motion and microscale interactions in diverse engineering systems.

Python Code Snippet

Below is a Python code snippet that encapsulates the core computational elements of vorticity dynamics including the calculation of vorticity, the vorticity transport equation, vortex stretching and tilting dynamics, and effects within turbulent flow applications.

```python
import numpy as np
from scipy.ndimage import gaussian_filter

def calculate_vorticity(velocity_field):
    '''
    Calculate the vorticity of a velocity field.
    :param velocity_field: A 2D or 3D array representing the
        velocity field.
    :return: Vorticity of the velocity field.
    '''
    dvx_dy = np.gradient(velocity_field[0], axis=1)
    dvy_dx = np.gradient(velocity_field[1], axis=0)
    vorticity = dvy_dx - dvx_dy
    return vorticity

def vorticity_transport(vorticity, velocity_field, nu, time_step,
    grid_spacing):
    '''
    Solves the vorticity transport equation.
    :param vorticity: Vorticity field.
    :param velocity_field: Velocity field consisting of (u, v)
        components.
    :param nu: Kinematic viscosity.
    :param time_step: Time step for simulation.
    :param grid_spacing: Spatial grid spacing.
    :return: Updated vorticity field.
    '''
    # Compute convection term
    conv_term = (
```

```python
        velocity_field[0] * np.gradient(vorticity, axis=0) +
        velocity_field[1] * np.gradient(vorticity, axis=1)
    )

    # Compute diffusion term
    laplacian_vorticity = (
        np.gradient(np.gradient(vorticity, axis=0), axis=0) +
        np.gradient(np.gradient(vorticity, axis=1), axis=1)
    )

    # Update vorticity using Euler method
    vorticity = vorticity + time_step * (-conv_term + nu *
    ↪ laplacian_vorticity)
    return vorticity

def vortex_stretching_tilting(vorticity, velocity_field):
    '''
    Analyze vortex stretching and tilting in a velocity field.
    :param vorticity: Vorticity field.
    :param velocity_field: Velocity field.
    :return: Stretching and tilting effects on vorticity.
    '''
    grad_v = np.gradient(velocity_field)
    stretching_term = vorticity * grad_v[0]  # Vortex stretching
    tilting_term = vorticity * grad_v[1]     # Vortex tilting
    return stretching_term, tilting_term

def apply_turbulent_effects(vorticity, velocity_field,
↪ eddy_viscosity):
    '''
    Incorporate turbulent effects into the vorticity field.
    :param vorticity: Vorticity field.
    :param velocity_field: Velocity field.
    :param eddy_viscosity: Effective viscosity for turbulence.
    :return: Adjusted vorticity field considering turbulence.
    '''
    smoothed_vorticity = gaussian_filter(vorticity,
    ↪ sigma=1/eddy_viscosity)
    return smoothed_vorticity

# Example velocity field (2D)
velocity_field = np.array([[[1, 2], [2, 4]], [[2, 4], [1, 2]]])
vorticity = calculate_vorticity(velocity_field)

# Parameters for vorticity transport
nu = 0.01
time_step = 0.1
grid_spacing = 1.0

# Update vorticity field
vorticity_updated = vorticity_transport(vorticity, velocity_field,
↪ nu, time_step, grid_spacing)
```

```
# Examine vortex stretching and tilting
stretching, tilting = vortex_stretching_tilting(vorticity,
↪    velocity_field)

# Apply turbulent effects
eddy_viscosity = 0.1
vorticity_with_turbulence =
↪    apply_turbulent_effects(vorticity_updated, velocity_field,
↪    eddy_viscosity)

print("Original Vorticity:", vorticity)
print("Updated Vorticity:", vorticity_updated)
print("Vortex Stretching:", stretching)
print("Vortex Tilting:", tilting)
print("Vorticity with Turbulence Effects:",
↪    vorticity_with_turbulence)
```

This code defines several key functions necessary for understanding and analyzing vorticity dynamics in fluid flows:

- `calculate_vorticity` function computes the vorticity from a given velocity field.

- `vorticity_transport` simulates the transport of vorticity using the vorticity transport equation.

- `vortex_stretching_tilting` identifies and analyzes the effects of vortex stretching and tilting in the fluid.

- `apply_turbulent_effects` adjusts the vorticity field to incorporate effects typically observed in turbulent flows using a smoothing technique.

The example at the end of the code demonstrates how to use these functions with a sample velocity field along with typical parameters used in fluid dynamics simulations.

Multiple Choice Questions

1. What physical quantity is vorticity mathematically associated with in fluid mechanics?

    (a) Divergence of the velocity field

    (b) Gradient of the pressure field

    (c) Curl of the velocity field

(d) Laplacian of the temperature field

2. How does vorticity conservation manifest in inviscid flows?

    (a) Vorticity is exponentially amplified along pathlines
    (b) Vorticity is conserved along pathlines
    (c) Vorticity diffuses isotropically
    (d) Vorticity is suppressed by external forces

3. Which term in the vorticity transport equation indicates viscous diffusion?

    (a) $\vec{v} \cdot \nabla \vec{\omega}$
    (b) $\vec{\omega} \cdot \nabla \vec{v}$
    (c) $\nu \nabla^2 \vec{\omega}$
    (d) $\frac{\partial \vec{\omega}}{\partial t}$

4. What does the circulation Γ in a fluid represent?

    (a) The potential energy per unit volume
    (b) The total momentum of the fluid system
    (c) The angular momentum around a closed contour
    (d) The closed integral of the velocity along a contour

5. In the context of vortex dynamics, what is vortex stretching primarily associated with?

    (a) Increase in pressure along the vortex line
    (b) Decrease of density in the vortex core
    (c) Increase in vorticity magnitude due to elongation
    (d) Expansion of the vortex core diameter

6. How do RANS methods relate to vorticity in turbulent flow modeling?

    (a) By ignoring vorticity effects to simplify models
    (b) By emphasizing vorticity in small-scale eddy simulations
    (c) By averaging pressure fields to account for vorticity
    (d) By incorporating vorticity interactions with mean flow

7. Which practical engineering application benefits significantly from understanding vorticity dynamics?

(a) Battery life optimization

(b) Structural load analysis

(c) Aerodynamic drag reduction

(d) Electric circuitry design

Answers:

1. **C: Curl of the velocity field** Vorticity is defined as the curl of the velocity field, which describes the local spinning motion of fluid elements.

2. **B: Vorticity is conserved along pathlines** In inviscid flows, vorticity is conserved along pathlines, reflecting the absence of viscous forces to alter vorticity magnitude.

3. **C: $\nu\nabla^2\vec{\omega}$** The term $\nu\nabla^2\vec{\omega}$ in the vorticity transport equation represents viscous diffusion, showing how vorticity spreads out over time due to viscosity.

4. **D: The closed integral of the velocity along a contour** Circulation Γ is defined as the closed loop integral of velocity around a given contour, representing the rotational flow within that loop.

5. **C: Increase in vorticity magnitude due to elongation** Vortex stretching occurs when a vortex line is extended, resulting in an increase in the vorticity magnitude, which is significant in turbulent flows.

6. **D: By incorporating vorticity interactions with mean flow** RANS methods average the equations of motion and consider the interaction between vorticity and the mean flow structures to capture turbulence effects.

7. **C: Aerodynamic drag reduction** In aerodynamic applications, understanding and controlling vorticity is essential to reduce drag and optimize lift, crucial for efficient aircraft design.

Chapter 12

K- (K-Epsilon) Turbulence Model

Introduction to Turbulence Modeling

The K-epsilon (k-ϵ) turbulence model is a semi-empirical model used for predicting turbulence in fluid dynamics. It is commonly implemented in solving practical engineering problems due to its robustness and simplicity for simulating complex turbulent flows. The model focuses on the transport equations for the turbulent kinetic energy, denoted by k, and its rate of dissipation, denoted by ϵ. The turbulence model provides closure to the Reynolds-Averaged Navier-Stokes (RANS) equations.

Turbulent Kinetic Energy Equation

The turbulent kinetic energy k quantifies the energy contained in the turbulent eddies per unit mass and is defined as:

$$k = \frac{1}{2}\left(\overline{u'_i u'_i}\right)$$

where $\overline{u'_i u'_i}$ is the velocity fluctuation covariance of the turbulent velocity components. The transport equation for k is derived from the RANS equations and is written as:

$$\frac{\partial k}{\partial t} + \overline{U_j}\frac{\partial k}{\partial x_j} = P_k - \epsilon + \frac{\partial}{\partial x_j}\left[\left(\nu + \frac{\nu_t}{\sigma_k}\right)\frac{\partial k}{\partial x_j}\right]$$

where P_k represents the production of turbulent kinetic energy, ϵ is the rate of dissipation of k, ν_t is the turbulent viscosity, and σ_k is the turbulent Prandtl number for k.

Dissipation Rate Equation

The dissipation rate ϵ characterizes the rate at which turbulent kinetic energy is converted into thermal internal energy and dissipated, primarily due to viscous effects. The transport equation for ϵ is given by:

$$\frac{\partial \epsilon}{\partial t} + \overline{U_j}\frac{\partial \epsilon}{\partial x_j} = C_1 \frac{\epsilon}{k} P_k - C_2 \frac{\epsilon^2}{k} + \frac{\partial}{\partial x_j}\left[\left(\nu + \frac{\nu_t}{\sigma_\epsilon}\right)\frac{\partial \epsilon}{\partial x_j}\right]$$

where C_1 and C_2 are empirical constants obtained through experimental data, and σ_ϵ is the turbulent Prandtl number for ϵ.

Closure Coefficients and Turbulent Viscosity

The calculation of the turbulent viscosity ν_t is critical in the K-epsilon model, and it is modeled as:

$$\nu_t = C_\mu \frac{k^2}{\epsilon}$$

where C_μ is another empirically determined constant. Typical values for the coefficients are $C_1 = 1.44$, $C_2 = 1.92$, $C_\mu = 0.09$, $\sigma_k = 1.0$, and $\sigma_\epsilon = 1.3$. These constants help guarantee that the model performs reliably across different flow conditions.

Application in Complex Fluid Systems

The K-epsilon model's application spans from simple geometries to complex systems, such as flow over airfoils, within combustion chambers, and around industrial structures. By employing the turbulence model within computational fluid dynamics (CFD), it allows for a more detailed prediction of flow separation, recirculation, and pressure loss. The ease of implementation paired with its capability to provide reasonable accuracy makes it an invaluable tool in many practical engineering scenarios.

Limitations and Opportunities for Improvement

Despite its widespread use, the `K-epsilon` model has limitations, especially in predicting near-wall turbulent flows and anisotropic turbulence. Opportunities for enhancements lie in the development of revised models, such as the `RNG K-epsilon` and `realizable K-epsilon`, which modify the base equations to capture a broader spectrum of turbulent phenomena and extend to more challenging flow conditions.

Python Code Snippet

Below is a Python code snippet that encompasses the core computational elements of the K-epsilon turbulence model, including the calculation of turbulent kinetic energy, dissipation rates, and necessary coefficients.

```python
import numpy as np

def turbulent_kinetic_energy(u_prime_i_rms):
    '''
    Calculate the turbulent kinetic energy (k) from RMS of velocity
      fluctuations.
    :param u_prime_i_rms: RMS of velocity fluctuations for component
      i.
    :return: Turbulent kinetic energy k.
    '''
    return 0.5 * np.sum(u_prime_i_rms ** 2)

def production_of_tke(U_j, grad_U_i, nu_t, sigma_k):
    '''
    Calculate the production term P_k for the turbulent kinetic
      energy equation.
    :param U_j: Mean velocity component.
    :param grad_U_i: Gradient of velocity component i.
    :param nu_t: Turbulent viscosity.
    :param sigma_k: Turbulent Prandtl number for k.
    :return: Production term P_k.
    '''
    return nu_t * np.sum(U_j * grad_U_i) / sigma_k

def dissipation_rate(k, epsilon, C1, P_k, C2):
    '''
    Calculate the dissipation rate epsilon for the dissipation rate
      equation.
```

```
    :param k: Turbulent kinetic energy.
    :param epsilon: Dissipation rate.
    :param C1: Empirical constant.
    :param P_k: Production term.
    :param C2: Empirical constant.
    :return: New dissipation rate.
    '''
    return C1 * epsilon * P_k / k - C2 * (epsilon ** 2) / k

def turbulent_viscosity(k, epsilon, C_mu):
    '''
    Calculate the turbulent viscosity nu_t.
    :param k: Turbulent kinetic energy.
    :param epsilon: Dissipation rate of k.
    :param C_mu: Empirical constant.
    :return: Turbulent viscosity nu_t.
    '''
    return C_mu * k ** 2 / epsilon

# Example usage with some dummy data
u_prime_i_rms = np.array([0.5, 0.5, 0.5])
U_j = np.array([1.0, 1.0, 1.0])
grad_U_i = np.array([0.1, 0.1, 0.1])
k = turbulent_kinetic_energy(u_prime_i_rms)
nu_t = turbulent_viscosity(k, 0.3, 0.09)
P_k = production_of_tke(U_j, grad_U_i, nu_t, 1.0)
epsilon = dissipation_rate(k, 0.3, 1.44, P_k, 1.92)

print("Turbulent Kinetic Energy (k):", k)
print("Turbulent Viscosity (nu_t):", nu_t)
print("Production Term (P_k):", P_k)
print("Dissipation Rate (epsilon):", epsilon)
```

This code defines several key functions necessary for simulating turbulence using the K-epsilon model:

- `turbulent_kinetic_energy` function computes the turbulent kinetic energy k from RMS of velocity fluctuations.

- `production_of_tke` calculates the production term P_k for the turbulent kinetic energy equation.

- `dissipation_rate` computes the dissipation rate ϵ based on energy turnover and dissipation via viscous effects.

- `turbulent_viscosity` calculates the turbulent viscosity ν_t using empirical constants.

The final section of the code demonstrates the calculation of these quantities using example data, illustrating the typical workflow when employing the K-epsilon model in fluid dynamics simulations.

Multiple Choice Questions

1. Turbulence modeling in fluid dynamics primarily aims to:
    (a) Reduce computational cost by neglecting fluid viscosity
    (b) Predict and simulate flow separation and aerodynamic drag
    (c) Solve the complete Navier-Stokes equations directly for all flow scales
    (d) Implement linear flow models for improved predictions

2. In the K-ϵ turbulence model, the dissipation rate ϵ is associated with:
    (a) The energy storage in turbulent eddies
    (b) The rate of production of turbulent kinetic energy
    (c) The conversion of turbulent kinetic energy into molecular internal energy
    (d) The transfer of energy between large-scale and small-scale structures

3. The turbulent kinetic energy k in the K-ϵ model is defined as:
    (a) The sum of potential and kinetic energy in the flow
    (b) The average kinetic energy per unit volume of turbulent eddies
    (c) Half the mean of the product of velocity fluctuations
    (d) The dissipation rate of kinetic energy due to viscosity

4. Typical value of the closure coefficient C_μ used in the calculation of turbulent viscosity in the K-ϵ model is:
    (a) 0.1
    (b) 0.2
    (c) 0.09
    (d) 1.44

5. How do the constants C_1 and C_2 in the ϵ equation influence the model?

   (a) They determine the stability of numerical implementations

   (b) They adjust the balance between turbulent energy production and dissipation

   (c) They control the level of turbulence in laminar flows

   (d) They modify the boundary layer thickness predictions

6. Which of the following is a limitation of the standard K-ϵ turbulence model?

   (a) Overestimation of pressure drag

   (b) Inability to predict shock wave interactions

   (c) Difficulty in accurately modeling near-wall turbulence

   (d) Complexity in implementation for simple geometries

7. What improvements do variations like the RNG and realizable K-ϵ models offer over the standard K-ϵ model?

   (a) Enhanced computational efficiency for laminar flows

   (b) Better prediction of isotropic turbulence

   (c) Improved capability for handling complex turbulent flows and near-wall regions

   (d) Simplified mathematical formulation for faster computation

Answers:

1. **B: Predict and simulate flow separation and aerodynamic drag** Turbulence modeling aims to accurately predict complex flow phenomena like separation and drag, which are crucial in many engineering applications.

2. **C: The conversion of turbulent kinetic energy into molecular internal energy** The parameter ϵ represents the rate at which turbulent kinetic energy is dissipated to thermal energy, often through viscous effects.

3. **C: Half the mean of the product of velocity fluctuations** The turbulent kinetic energy is defined as half the mean of the squared velocity fluctuations, capturing the energy in eddies.

4. **C: 0.09** The closure coefficient C_μ is empirically determined and typically set to 0.09 in the standard K-ϵ model for predicting turbulent viscosity.

5. **B: They adjust the balance between turbulent energy production and dissipation** The constants C_1 and C_2 help balance production and dissipation rates, influencing turbulence characteristics in simulations.

6. **C: Difficulty in accurately modeling near-wall turbulence** The K-ϵ model struggles with near-wall turbulence, requiring modifications or additional models to improve predictions.

7. **C: Improved capability for handling complex turbulent flows and near-wall regions** Variants like RNG and realizable K-ϵ models provide enhancements in terms of accuracy for complex and near-wall turbulence, addressing some limitations of the standard model.

Chapter 13

K- (K-Omega) Turbulence Model

The K- Model Formulation

The K-omega (k-ω) turbulence model represents a significant development in turbulence modeling, particularly beneficial for predicting near-wall flows with improved accuracy over the K-epsilon model. It provides closure to the Reynolds-Averaged Navier-Stokes (RANS) equations using two specific transport equations for the turbulence kinetic energy k and the specific dissipation rate ω.

The turbulent kinetic energy equation is given by:

$$\frac{\partial k}{\partial t} + U_j \frac{\partial k}{\partial x_j} = P_k - \beta^* k\omega + \frac{\partial}{\partial x_j}\left[(\nu + \sigma_k^* \nu_t)\frac{\partial k}{\partial x_j}\right]$$

where: - P_k is the production term of turbulent kinetic energy, - β^* is an empirical constant related to the dissipation of k, - σ_k^* is the turbulent Prandtl number for k.

The specific dissipation rate, ω, is associated with the scale of the turbulence. Its transport equation is given by:

$$\frac{\partial \omega}{\partial t} + U_j \frac{\partial \omega}{\partial x_j} = \frac{\gamma}{\nu_t} P_k - \beta \omega^2 + \frac{\partial}{\partial x_j}\left[(\nu + \sigma_\omega \nu_t)\frac{\partial \omega}{\partial x_j}\right] + 2(1-F_1)\frac{\sigma_{\omega 2}}{\omega}\frac{\partial k}{\partial x_j}\frac{\partial \omega}{\partial x_j}$$

where: - γ, β, σ_ω, and $\sigma_{\omega 2}$ are empirical constants, - F_1 is a blending function.

Near-Wall Turbulence Modeling

The K-omega model's primary advantage is its ability to accurately predict flow in the near-wall region without additional wall functions. It excelled due to the specific dissipation rate ω, which provides more physical representations of flow characteristics close to the wall.

The near-wall treatment is delineated by the blending function F_1, ensuring a smooth transition between the near-wall and free-stream regions. The definition of F_1 is integral, optimizing performance across various flow scenarios.

$$F_1 = \tanh\left(\min\left[\max\left(\frac{\sqrt{k}}{\beta^*\omega y}, \frac{500\nu}{y^2\omega}\right), \frac{4\sigma_\omega 2k}{CD^* k\omega y^2}\right]\right)$$

where y is the distance to the nearest wall, and CD^* is a constant for better adaptation to the wake region.

Empirical Constants and Turbulent Viscosity

Empirical constants in the K-omega model ensure its robustness across diverse applications. Typical values are $\beta^* = 0.09$, $\beta = 0.072$, $\sigma_k = 0.5$, $\sigma_\omega = 0.5$, and γ often calibrated for specific applications.

The model calculates the turbulent eddy viscosity ν_t using:

$$\nu_t = \frac{k}{\omega}$$

This formulation highlights the role of ω in seamlessly incorporating the effects of turbulence scales, leading to reliable predictions, especially around boundaries.

Application and Numerical Stability

The application of the K-omega model covers a wide range of engineering scenarios, including but not limited to, aerodynamic profiles, combustor design, and the analysis of flow separation. The model's inherent advantage in handling near-wall turbulence without empirical wall functions enhances its stability in complex geometries.

Numerical stability is often a concern in simulations involving complex geometries or highly anisotropic flows. The K-omega model addresses these issues through its formulation, providing consistent and robust results when adequately implemented within computational fluid dynamics frameworks.

Python Code Snippet

Below is a Python code snippet that encompasses the core computational elements associated with the K-omega turbulence model, including the calculation of turbulent kinetic energy, specific dissipation rate, empirical constants, and their application in numerical stability assessments.

```
import numpy as np

def turbulent_kinetic_energy(U_j, P_k, beta_star, k, omega, nu,
    sigma_k_star, grad_k):
    '''
    Calculate the turbulent kinetic energy.
    :param U_j: Velocity vector component.
    :param P_k: Production term.
    :param beta_star: Empirical constant.
    :param k: Turbulence kinetic energy.
    :param omega: Specific dissipation rate.
    :param nu: Kinematic viscosity.
    :param sigma_k_star: Turbulent Prandtl number for k.
    :param grad_k: Gradient of k.
    :return: Change in turbulent kinetic energy.
    '''
    return - U_j * np.dot(grad_k, U_j) + P_k - beta_star * k * omega
        + np.dot((nu + sigma_k_star), grad_k)

def specific_dissipation_rate(U_j, gamma, nu_t, P_k, beta, omega,
    nu, sigma_omega, sigma_omega2, grad_k, grad_omega, F_1):
    '''
    Calculate the specific dissipation rate.
    :param U_j: Velocity vector component.
    :param gamma: Empirical constant.
    :param nu_t: Turbulent eddy viscosity.
    :param P_k: Production term.
    :param beta: Empirical constant.
    :param omega: Specific dissipation rate.
    :param nu: Kinematic viscosity.
```

```python
    :param sigma_omega: Turbulent Prandtl number for omega.
    :param sigma_omega2: Empirical constant.
    :param grad_k: Gradient of k.
    :param grad_omega: Gradient of omega.
    :param F_1: Blending function.
    :return: Change in specific dissipation rate.
    '''
    return (
        - U_j * np.dot(grad_omega, U_j)
        + (gamma / nu_t) * P_k
        - beta * omega ** 2
        + np.dot((nu + sigma_omega), grad_omega)
        + 2 * (1 - F_1) * (sigma_omega2 / omega) * np.dot(grad_k,
        ↪    grad_omega)
    )

def blending_function(y, k, omega, nu, beta_star, sigma_omega2,
↪    CD_star):
    '''
    Calculate the blending function F_1.
    :param y: Distance to nearest wall.
    :param k: Turbulence kinetic energy.
    :param omega: Specific dissipation rate.
    :param nu: Kinematic viscosity.
    :param beta_star: Empirical constant.
    :param sigma_omega2: Empirical constant.
    :param CD_star: Constant for wake region.
    :return: Value of blending function F_1.
    '''
    return np.tanh(
        np.min([
            np.max([np.sqrt(k) / (beta_star * omega * y), 500 * nu /
            ↪    (y ** 2 * omega)]),
            4 * sigma_omega2 * k / (CD_star * k * omega * y ** 2)
        ])
    )

def turbulent_eddy_viscosity(k, omega):
    '''
    Calculate the turbulent eddy viscosity.
    :param k: Turbulence kinetic energy.
    :param omega: Specific dissipation rate.
    :return: Turbulent eddy viscosity.
    '''
    return k / omega

# Constants for demonstration purposes
beta_star = 0.09
beta = 0.072
sigma_k = 0.5
sigma_omega = 0.5
gamma = 0.6  # This value should be calibrated within the context of
↪    application
```

```python
# Example usage of the functions
U_j = np.array([1.0, 2.0, 3.0]) # Velocity vector example
P_k = 1.0 # Example production term
k = 0.5 # Example turbulence kinetic energy
omega = 0.4 # Example specific dissipation rate
nu = 1e-5 # Example kinematic viscosity
sigma_k_star = 0.5
sigma_omega2 = 0.2
grad_k = np.array([0.1, 0.2, 0.3]) # Gradient of k
grad_omega = np.array([0.1, 0.2, 0.3]) # Gradient of omega
y = 0.05 # Distance to the nearest wall
CD_star = 1.0 # Example constant for blending function

change_tke = turbulent_kinetic_energy(U_j, P_k, beta_star, k, omega,
    nu, sigma_k_star, grad_k)
change_omega = specific_dissipation_rate(U_j, gamma,
    turbulent_eddy_viscosity(k, omega), P_k, beta, omega, nu,
    sigma_omega, sigma_omega2, grad_k, grad_omega,
    blending_function(y, k, omega, nu, beta_star, sigma_omega2,
    CD_star))
eddy_viscosity = turbulent_eddy_viscosity(k, omega)
F_1 = blending_function(y, k, omega, nu, beta_star, sigma_omega2,
    CD_star)

print("Change in Turbulent Kinetic Energy:", change_tke)
print("Change in Specific Dissipation Rate:", change_omega)
print("Turbulent Eddy Viscosity:", eddy_viscosity)
print("Blending Function F_1:", F_1)
```

This code defines several key functions necessary for implementing the K-omega model in computational simulations:

- `turbulent_kinetic_energy` computes the changes in turbulent kinetic energy based on flow conditions.

- `specific_dissipation_rate` calculates changes in the specific dissipation rate, which affects turbulence scales.

- `blending_function` provides the transition function between near-wall and free-stream flows.

- `turbulent_eddy_viscosity` computes the eddy viscosity, essential for resolving turbulent flows.

The final block demonstrates the application of these functions using exemplary data. This setup helps in understanding and simulating complex fluid dynamics scenarios, especially when considering near-wall flow characteristics.

Multiple Choice Questions

1. Which of the following is a primary advantage of the K-omega (k-ω) turbulence model?

    (a) Simplifies calculations in the free-stream region
    (b) Provides enhanced predictions for near-wall flows
    (c) Reduces computational cost in all flow regimes
    (d) Eliminates the need for empirical constants

2. The specific dissipation rate ω in the K-omega model is primarily associated with:

    (a) Energy production in turbulent flows
    (b) The scale of turbulence
    (c) Momentum transfer in laminar flows
    (d) Compressibility effects in high-speed flows

3. In the K-omega model, which term in the turbulent kinetic energy equation represents energy dissipation?

    (a) P_k
    (b) $\beta^* k\omega$
    (c) $\sigma_k^* \nu_t$
    (d) $\frac{\partial}{\partial x_j}$

4. What role does the blending function F_1 play in the K-omega turbulence model?

    (a) It determines the inclusion of compressibility effects
    (b) It enhances the transition between laminar and turbulent flows
    (c) It governs the transition between near-wall and free-stream regions
    (d) It adjusts the empirical constants based on flow conditions

5. Which of the following is NOT a typical empirical constant found in the K-omega model?

    (a) β^*

(b) σ_k

   (c) γ^*

   (d) σ_ω

6. How is turbulent viscosity ν_t calculated in the K-omega model?

   (a) $\nu_t = \omega^2/k$

   (b) $\nu_t = \beta\omega/k$

   (c) $\nu_t = \gamma k/\omega$

   (d) $\nu_t = k/\omega$

7. In which applications is the K-omega model particularly advantageous?

   (a) High-speed car aerodynamics

   (b) Combustion chamber design

   (c) Flow separation analysis

   (d) All of the above

Answers:

1. **B: Provides enhanced predictions for near-wall flows** The K-omega model excels in predicting near-wall flows due to the specific dissipation rate ω, offering better accuracy than the K-epsilon model without requiring wall functions.

2. **B: The scale of turbulence** The specific dissipation rate ω directly relates to the scale or frequency of turbulence, making it critical in predicting how turbulence diminishes at various scales.

3. **B: $\beta^* k\omega$** This term in the turbulent kinetic energy equation accounts for the dissipation of turbulent kinetic energy within the flow, a crucial aspect of turbulence modeling.

4. **C: It governs the transition between near-wall and free-stream regions** The blending function F_1 ensures smooth transitioning in the model's application both near walls and in the free stream, optimizing performance across different flow scenarios.

5. **C: γ^*** γ^* is not typically used as an empirical constant in the K-omega model. The constants used include β^*, σ_k, and σ_ω.

6. **D: $\nu_t = k/\omega$** The turbulent viscosity is computed using this ratio, which incorporates the effects of both turbulent kinetic energy and specific dissipation rate to model turbulence scales.

7. **D: All of the above** The K-omega model is beneficial in a wide range of applications, from aerodynamics and combustion to analyzing flow separation, due to its robust handling of different turbulence characteristics.

Chapter 14

Momentum Integral Equation

Introduction to Momentum Integral Equation

The momentum integral equation is a pivotal tool in fluid mechanics for analyzing boundary layers. It is derived by integrating the differential momentum equation across the boundary layer. This approach simplifies the complex flow analysis near solid boundaries without solving the full Navier-Stokes equations.

The general form of the momentum integral equation can be expressed as follows:

$$\frac{d}{dx}(\delta^* U_\infty) + \theta \frac{dU_\infty}{dx} = C_f$$

where δ^* is the displacement thickness, θ is the momentum thickness, U_∞ is the free stream velocity, and C_f is the skin friction coefficient.

Displacement and Momentum Thickness

Displacement thickness δ^* is a measure of the boundary layer's effect on the external flow, defined as:

$$\delta^* = \int_0^\infty \left(1 - \frac{u}{U_\infty}\right) dy$$

where u is the velocity parallel to the wall within the boundary layer.

Momentum thickness θ quantifies the momentum deficit in the boundary layer:

$$\theta = \int_0^\infty \frac{u}{U_\infty}\left(1 - \frac{u}{U_\infty}\right) dy$$

Both thicknesses are crucial in predicting the boundary layer characteristics and their interactions with solid surfaces.

Derivation of the Momentum Integral Equation

The derivation of the momentum integral equation starts from the differential form of the momentum equation in the boundary layer, which is given by:

$$\rho\left(u\frac{\partial u}{\partial x} + v\frac{\partial u}{\partial y}\right) = -\frac{\partial p}{\partial x} + \mu\frac{\partial^2 u}{\partial y^2}$$

Integrating the above equation across the boundary layer from $y = 0$ to $y = \infty$ and applying suitable assumptions leads to the momentum integral form. Key assumptions include neglecting the pressure variations in the y-direction (i.e., $\partial p/\partial y = 0$).

Application in Boundary Layer Prediction

In practice, the momentum integral equation is often employed to predict boundary layer behavior on flat and curved surfaces. The equation provides insights into skin friction drag over surfaces and aids in determining separation points.

Using the Thwaites' correlation for laminar flow, the momentum equation becomes a powerful predictive tool:

$$C_f = 0.664\sqrt{\frac{\nu}{U_\infty x}}$$

This empirical relation links the shear stress at the wall to the velocity and kinematic viscosity of the fluid, aiding in design predictions for fluid flow over aircraft wings, automobiles, and other structures in engineering.

1 Utilization in Laminar and Turbulent Flows

For laminar flows, the momentum integral equation is simplified and often yields analytical solutions. However, in turbulent flows, empirical methods or approximations, like the Von Kármán momentum integral for turbulent boundary layers, are necessary:

$$\frac{\delta^*}{\theta} = \frac{\kappa}{2} \frac{C_f}{2}$$

where κ represents the Kármán constant.

Numerical Methods and Computational Applications

Numerical approaches solve the momentum integral equation in complex geometries where analytical solutions aren't feasible. Computational Fluid Dynamics (CFD) often involves iterative methods to approximate δ^*, θ, and C_f under varying conditions.

The RANS (Reynolds-Averaged Navier-Stokes) equations frequently incorporate the momentum integral approach to bridge the gap between theoretical and real-world applications, particularly in scenarios involving turbulence modeling and transition predictions in boundary layers.

Python Code Snippet

Below is a Python code snippet that encompasses the core computational elements related to the momentum integral equation, including the calculation of displacement and momentum thickness, and an example of how these are used in practice.

```
import numpy as np
from scipy.integrate import quad

def displacement_thickness(u, U_inf, y):
    '''
    Calculate the displacement thickness (*).
    :param u: Velocity profile as a function of y.
    :param U_inf: Free stream velocity.
    :param y: Grid or array of y values.
    :return: Displacement thickness.
    '''
    integrand = lambda y: 1 - u(y) / U_inf
```

```
        delta_star, _ = quad(integrand, y[0], y[-1])

        return delta_star

def momentum_thickness(u, U_inf, y):
    '''
    Calculate the momentum thickness ().
    :param u: Velocity profile as a function of y.
    :param U_inf: Free stream velocity.
    :param y: Grid or array of y values.
    :return: Momentum thickness.
    '''
    integrand = lambda y: (u(y) / U_inf) * (1 - u(y) / U_inf)
    theta, _ = quad(integrand, y[0], y[-1])

    return theta

def skin_friction_coefficient(U_inf, x, nu):
    '''
    Calculate the skin friction coefficient (C_f) using the Thwaites
    ↪ method for laminar flow.
    :param U_inf: Free stream velocity.
    :param x: Streamwise location or distance from the leading edge.
    :param nu: Kinematic viscosity of the fluid.
    :return: Skin friction coefficient.
    '''
    return 0.664 * np.sqrt(nu / (U_inf * x))

# Example velocity profile as a function of y (e.g., linear for
↪ simplicity)
velocity_profile = lambda y: 1.25 * y * U_inf if y <= 1 else U_inf

# Define parameters
U_inf = 1.0  # Assumed constant free stream velocity
y_values = np.linspace(0, 5, 100)  # Grid for integration
nu = 1.5e-5  # Kinematic viscosity in m^2/s
x_location = 0.5  # Streamwise location in meters

# Calculate displacement thickness, momentum thickness, and skin
↪ friction coefficient
delta_star = displacement_thickness(velocity_profile, U_inf,
↪ y_values)
theta = momentum_thickness(velocity_profile, U_inf, y_values)
C_f = skin_friction_coefficient(U_inf, x_location, nu)

print("Displacement Thickness (*):", delta_star)
print("Momentum Thickness ():", theta)
print("Skin Friction Coefficient (C_f):", C_f)
```

This code defines several key functions necessary for analyzing boundary layers using the momentum integral equation:

- `displacement_thickness` computes the displacement thickness δ^*, which reflects how the presence of the boundary layer displaces the potential flow ahead of it.

- `momentum_thickness` calculates the momentum thickness θ, indicating the reduction in momentum due to the boundary layer.

- `skin_friction_coefficient` estimates the skin friction coefficient C_f, using the Thwaites method for laminar flows, important for predicting drag over surfaces.

This code provides foundational tools necessary for solving fluid flow problems involving boundary layers, helping engineers understand and predict fluid behaviors near solid surfaces.

Multiple Choice Questions

1. What is the main advantage of utilizing the momentum integral equation in boundary layer analysis?

    (a) It provides an exact solution of Navier-Stokes equations.

    (b) It simplifies the analysis by integrating the differential momentum equation.

    (c) It directly computes turbulent flow characteristics.

    (d) It eliminates the need for empirical data in flow predictions.

2. Which thickness is considered when calculating the effect of a boundary layer on external flow?

    (a) Energy thickness

    (b) Displacement thickness

    (c) Laminar thickness

    (d) Viscous thickness

3. The displacement thickness, δ^*, is defined as:

    (a) The distance required for pressure recovery in a boundary layer.

    (b) The measure of the boundary layer's effect on the external flow.

- (c) The actual thickness of the viscous sublayer.
- (d) The height at which velocity becomes constant outside the boundary layer.

4. In the given momentum integral equation, which variable represents the skin friction coefficient?
   - (a) δ^*
   - (b) θ
   - (c) U_∞
   - (d) C_f

5. Which empirical relation is used in predicting skin friction coefficient for laminar flow over flat plates?
   - (a) Newton's law of viscosity
   - (b) Von Kármán relation
   - (c) Thwaites' correlation
   - (d) Prandtl's mixing-length theory

6. What assumption is made about pressure variations when deriving the momentum integral equation?
   - (a) Pressure varies linearly in the x-direction.
   - (b) Pressure is constant throughout the boundary layer.
   - (c) Pressure only varies significantly in the y-direction.
   - (d) $\partial p/\partial y = 0$, indicating negligible pressure variations in the y-direction.

7. Numerical methods in CFD typically solve the momentum integral equation for which purpose?
   - (a) To directly solve Navier-Stokes equations.
   - (b) To assist in boundary layer separation predictions.
   - (c) To replicate exact analytical solutions.
   - (d) To avoid any empirical correlations or approximations.

Answers:
1. **B: It simplifies the analysis by integrating the differential momentum equation** The momentum integral equation is derived by integrating the momentum equations, which simplifies

boundary layer analysis without the need for solving the complex full Navier-Stokes equations.

2. **B: Displacement thickness** Displacement thickness is a measure of how much the boundary layer displaces the external flow within the fluid body.

3. **B: The measure of the boundary layer's effect on the external flow** Displacement thickness represents the distance by which the outer potential flow is displaced due to the boundary layer's presence.

4. **D:** C_f In the momentum integral equation, C_f denotes the skin friction coefficient, representing the drag due to shear stress at the surface.

5. **C: Thwaites' correlation** Thwaites' correlation provides a method for estimating the skin friction coefficient for laminar boundary layers over flat surfaces and is frequently used in design applications.

6. **D:** $\partial p/\partial y = 0$**, indicating negligible pressure variations in the y-direction** This assumption simplifies the derivation of the momentum integral equation by neglecting changes in pressure normal to the wall.

7. **B: To assist in boundary layer separation predictions** Numerical methods and CFD use the momentum integral equation to predict boundary layer characteristics, such as separation and transition, when analytical solutions are not possible.

Chapter 15

Cavitating Flow Dynamics

Introduction to Cavitation Phenomena

Cavitation is an intricate phenomenon occurring in fluid flows when local pressure drops below the fluid's vapor pressure, leading to vapor bubble formation. Understanding cavitating flow dynamics is essential for the design and analysis of hydraulic systems, as cavitation can lead to material erosion, noise, and vibration.

Governing Equations in Cavitating Flows

The study of cavitating flows necessitates a comprehensive understanding of the governing equations, typically derived from the Navier-Stokes equations coupled with vapor-liquid phase change dynamics. The equations of motion for cavitating flow can be expressed using the continuity equation and the momentum equation.

The continuity equation for an incompressible fluid with phase change is given by:

$$\frac{\partial \rho}{\partial t} + \nabla \cdot (\rho \mathbf{u}) = \dot{m}$$

where ρ is the fluid density, $\mathbf{u}$ is the velocity vector, and $\dot{m}$ represents the mass transfer rate due to phase change.

The momentum equation can be expressed as:

$$\frac{\partial(\rho\mathbf{u})}{\partial t} + \nabla \cdot (\rho\mathbf{uu}) = -\nabla p + \nabla \cdot \boldsymbol{\tau} + \rho\mathbf{g}$$

where p denotes the pressure, $\boldsymbol{\tau}$ is the viscous stress tensor, and $\mathbf{g}$ is the gravitational acceleration vector.

Vapor Bubble Dynamics

The Rayleigh-Plesset equation is pivotal in modeling individual bubble dynamics within a cavitating flow, expressed as:

$$R\frac{d^2 R}{dt^2} + \frac{3}{2}\left(\frac{dR}{dt}\right)^2 = \frac{p_g - p_\infty - 4\mu \frac{dR}{dt}/R}{\rho}$$

where R is the bubble radius, p_g is the gas pressure inside the bubble, p_∞ is the pressure of the liquid far from the bubble, and μ is the dynamic viscosity.

1 Nucleation and Growth

The inception of cavitation includes nucleation sites and subsequent growth of vapor bubbles. Nucleation can be homogeneous or heterogeneous, often initiated at pre-existing microcavities or impurities.

The growth rate of cavitation bubbles can be described through the pressure difference and surface tension effects:

$$\frac{dR}{dt} = \sqrt{\frac{2(p_v - p)}{\rho_l}}$$

where p_v is the vapor pressure and ρ_l is the liquid density.

Implications in Hydraulic Systems

Cavitation significantly impacts hydraulic systems, including pumps, turbines, and propellers. The flow-induced instabilities can cause efficiency losses and mechanical damage. For instance, in centrifugal pumps, the NPSH (Net Positive Suction Head) is crucial to determining cavitation inception conditions:

$$\text{NPSH}_{\text{available}} = \frac{p_a + p_h - p_v}{\rho g}$$

where p_a is the atmospheric pressure, p_h is the static head, and g is the acceleration due to gravity.

Vorticity and flow separation due to cavitation must be analyzed in detail to prevent potential structural failures in hydraulic systems.

1 Experimental and Computational Studies

Numerical simulations, often employing RANS or LES models, are instrumental in predicting cavitating flow patterns. Experiments provide validation for these models, which include cavitation number (σ) as a non-dimensional parameter crucial to such studies:

$$\sigma = \frac{p_\infty - p_v}{0.5\rho U^2}$$

where U is the reference velocity of the flow.

Advanced CFD simulations and cavitation models facilitate a deeper understanding of complex cavitation behavior, enabling more accurate predictions of hydraulic machinery performance and lifetime.

Python Code Snippet

Below is a Python code snippet that encompasses the core computational elements for studying cavitating flow dynamics, including the calculation of nucleation rates, bubble growth, and evaluating cavitation conditions in hydraulic systems.

```python
import numpy as np

def cavitation_nucleation(p_inf, p_v, rho_l, sigma):
    '''
    Calculate the bubble growth rate due to pressure difference.
    :param p_inf: Pressure in the liquid far from the bubble.
    :param p_v: Vapor pressure.
    :param rho_l: Liquid density.
    :param sigma: Surface tension.
    :return: Growth rate.
    '''
    return np.sqrt(2 * (p_v - p_inf) / rho_l)

def bubble_dynamics_equation(R, p_g, p_inf, mu, rho):
    '''
    Solve Rayleigh-Plesset equation for bubble dynamics.
    :param R: Initial bubble radius.
```

```
    :param p_g: Gas pressure inside the bubble.
    :param p_inf: Pressure of the liquid far from the bubble.
    :param mu: Dynamic viscosity.
    :param rho: Fluid density.
    :return: Second derivative of the bubble radius.
    '''
    # Use a finite difference method for solving the second
    ↪  derivative
    dR_dt = 0.001  # A small time step increment
    R_t1 = R + dR_dt
    term1 = R_t1 * (R_t1 - 2 * R)
    term2 = 3/2 * np.power(dR_dt, 2)
    term3 = (p_g - p_inf - 4 * mu * dR_dt / R) / rho

    return term1 + term2 - term3

def npsh_available(p_a, p_h, p_v, rho, g):
    '''
    Calculate the Net Positive Suction Head (NPSH) available.
    :param p_a: Atmospheric pressure.
    :param p_h: Static head.
    :param p_v: Vapor pressure.
    :param rho: Fluid density.
    :param g: Acceleration due to gravity.
    :return: NPSH available.
    '''
    return (p_a + p_h - p_v) / (rho * g)

def cavitation_number(p_inf, p_v, rho, U):
    '''
    Calculate the cavitation number.
    :param p_inf: Pressure in the fluid far field.
    :param p_v: Vapor pressure.
    :param rho: Fluid density.
    :param U: Reference velocity of the flow.
    :return: Cavitation number.
    '''
    return (p_inf - p_v) / (0.5 * rho * U ** 2)

# Example of computation for cavitating conditions
p_inf_example = 101325  # Pa, atmospheric pressure
p_v_example = 2330  # Pa, typical vapor pressure of water
rho_l_example = 997  # kg/m^3, density of water
sigma_example = 0.072  # N/m, surface tension of water
g_example = 9.81  # m/s^2, acceleration due to gravity

# Calculate bubble growth rate
growth_rate = cavitation_nucleation(p_inf_example, p_v_example,
    ↪  rho_l_example, sigma_example)

# Evaluate bubble dynamics using Rayleigh-Plesset equation
R_example = 0.0001  # m, initial bubble radius
p_g_example = 101325  # Pa, typical gas pressure
```

```
mu_example = 0.001    # Pa.s, dynamic viscosity for water
bubble_dynamics = bubble_dynamics_equation(R_example, p_g_example,
↪    p_inf_example, mu_example, rho_l_example)

# Calculate NPSH available
p_a_example = 101325    # Pa, atmospheric pressure
p_h_example = 10    # m, static head
npsh = npsh_available(p_a_example, p_h_example, p_v_example,
↪    rho_l_example, g_example)

# Compute cavitation number
U_example = 2    # m/s, reference velocity
cavitation_num = cavitation_number(p_inf_example, p_v_example,
↪    rho_l_example, U_example)

print("Bubble Growth Rate:", growth_rate)
print("Bubble Dynamics:", bubble_dynamics)
print("NPSH Available:", npsh)
print("Cavitation Number:", cavitation_num)
```

This code defines several key functions necessary for the study and evaluation of cavitating flows:

- `cavitation_nucleation` calculates the rate of bubble growth based on pressure differences.

- `bubble_dynamics_equation` solves the Rayleigh-Plesset equation to study bubble dynamics under specific conditions.

- `npsh_available` computes the net positive suction head available, critical in avoiding cavitation in hydraulic applications.

- `cavitation_number` evaluates the cavitation number to assess the likelihood of cavitation occurring in a flow.

The final code block provides examples of calculating these core elements using typical fluid properties and conditions.

Multiple Choice Questions

1. Cavitation occurs in a fluid flow when:

   (a) The fluid velocity exceeds the speed of sound

   (b) Local pressure drops below the fluid's vapor pressure

   (c) The flow becomes turbulent

   (d) The fluid density increases dramatically

2. Which equation is crucial for modeling the dynamics of individual bubbles in cavitating flows?

    (a) Bernoulli's Equation

    (b) Continuity Equation

    (c) Rayleigh-Plesset Equation

    (d) Euler's Equation

3. In the context of cavitation, the mass transfer rate due to phase change in the continuity equation is represented by:

    (a) $\dot{p}$ (pressure change rate)

    (b) $\dot{q}$ (heat transfer rate)

    (c) $\dot{m}$ (mass transfer rate)

    (d) $\dot{v}$ (volume transfer rate)

4. The term "NPSH" stands for:

    (a) Net Pressure Suction Head

    (b) Net Positive Suction Head

    (c) Neutral Pressure Siphon Head

    (d) Negative Pressure Suction Height

5. What does the cavitation number (σ) compare in flow studies?

    (a) Pressure difference related to the fluid's vapor pressure with inertial forces

    (b) Density changes with gravitational forces

    (c) Viscous forces with thermal forces

    (d) Velocity potential with boundary layer thickness

6. Which phenomenon does cavitation lead to that can be detrimental to hydraulic systems?

    (a) Increased heat transfer

    (b) Enhanced lubrication

    (c) Material erosion

    (d) Acoustic shielding

7. Which method is often used to simulate cavitating flow patterns numerically?

   (a) CFD with RANS or LES models
   (b) Finite element analysis
   (c) Multibody dynamics
   (d) Boundary element method

Answers:

1. **B: Local pressure drops below the fluid's vapor pressure** Cavitation occurs when the local pressure in a fluid drops below the vapor pressure, causing bubbles to form.

2. **C: Rayleigh-Plesset Equation** The Rayleigh-Plesset equation is crucial for describing the dynamics and behavior of individual cavitation bubbles.

3. **C: $\dot{m}$ (mass transfer rate)** In cavitating flows, $\dot{m}$ represents the mass transfer rate associated with the phase change from liquid to vapor and vice versa.

4. **B: Net Positive Suction Head** NPSH is a crucial factor in pump design that determines the cavitation inception conditions by ensuring that enough suction head is available to prevent cavitation.

5. **A: Pressure difference related to the fluid's vapor pressure with inertial forces** The cavitation number (σ) calculates the balance between pressure forces and inertial forces to characterize cavitation potential.

6. **C: Material erosion** Cavitation can lead to material erosion due to the intense pressure changes when vapor bubbles collapse, damaging hydraulic components.

7. **A: CFD with RANS or LES models** Computational Fluid Dynamics (CFD) utilizing RANS or LES models is often used to simulate cavitating flow patterns, providing detailed flow insights.

Chapter 16

Multiphase Flow Quantification

Governing Equations for Multiphase Flow

Multiphase flows are characterized by the simultaneous presence of different phases, such as gas, liquid, and solid, in a complex interaction within a system. The Navier-Stokes equations form the foundation but require adaptation to capture the dynamics of each phase accurately. The multiphase flow models often deal with continuity, momentum, and energy conservation equations for each phase.

1 Continuity Equation

The continuity equation for the k-th phase in a multiphase system is expressed as:

$$\frac{\partial(\alpha_k \rho_k)}{\partial t} + \nabla \cdot (\alpha_k \rho_k \mathbf{u}_k) = \Gamma_k$$

where α_k is the volume fraction, ρ_k is the density, $\mathbf{u}_k$ is the velocity vector of the k-th phase, and Γ_k represents the inter-phase mass transfer.

2 Momentum Equation

The momentum equation for each phase is critical in capturing the momentum exchange between phases and is formulated as:

$$\frac{\partial(\alpha_k \rho_k \mathbf{u}_k)}{\partial t} + \nabla \cdot (\alpha_k \rho_k \mathbf{u}_k \mathbf{u}_k) = \alpha_k \rho_k \mathbf{g} - \alpha_k \nabla p + \nabla \cdot \boldsymbol{\tau}_k + \mathbf{F}_{kl}$$

where p denotes the pressure, $\boldsymbol{\tau}_k$ is the viscous stress tensor, and $\mathbf{F}_{kl}$ represents the inter-phase momentum exchange term.

3 Energy Equation

Energy equations account for the thermal interactions between phases. For the k-th phase, the equation can be described as:

$$\frac{\partial(\alpha_k \rho_k E_k)}{\partial t} + \nabla \cdot (\alpha_k \rho_k E_k \mathbf{u}_k) = -\nabla \cdot \mathbf{q}_k + \alpha_k \rho_k \mathbf{u}_k \cdot \mathbf{g} + \Phi_k + \dot{Q}_{kl}$$

where E_k is the specific total energy, $\mathbf{q}_k$ the heat flux, Φ_k the viscous dissipation, and $\dot{Q}_{kl}$ the inter-phase heat transfer.

Modeling Strategies for Multiphase Flows

The modeling of multiphase flows involves simplifying assumptions and empirical relations to balance computational cost with physical accuracy in complex geometries.

1 Volume of Fluid (VOF) Method

The Volume of Fluid (VOF) method is a surface-tracking technique used in fixed meshes, suitable for tracking the interface between immiscible fluids. The interface is represented by a scalar function C:

$$\frac{\partial C}{\partial t} + \mathbf{u} \cdot \nabla C = 0$$

where C ranges from 0 to 1, identifying the phases within the domain.

2 Eulerian-Eulerian Models

These models treat each phase as interpenetrating continua, employing balance equations for volume fractions, momentum, and energy. A critical aspect is the inclusion of drag models for momentum exchange:

$$\mathbf{F}_{kl}^{\text{drag}} = \frac{3}{4} \frac{C_D \rho_l \alpha_k \alpha_l |\mathbf{u}_k - \mathbf{u}_l|(\mathbf{u}_k - \mathbf{u}_l)}{d_{kl}}$$

where C_D is the drag coefficient, ρ_l is the liquid density, and d_{kl} is the characteristic length scale.

3 Lagrangian Approaches

Lagrangian methods track individual particles or bubbles, providing detailed insights into particle-fluid interactions. The motion of these discrete elements is governed by Newton's second law:

$$m_p \frac{d\mathbf{u}_p}{dt} = \sum \mathbf{F}_p$$

where m_p is the particle mass and $\mathbf{F}_p$ are the forces acting on the particle, including drag, lift, and buoyancy.

Simulation Techniques

Numerical simulation of multiphase systems requires robust algorithms and computing power to solve the complex set of equations governing such flows.

1 Finite Volume Method (FVM)

FVM is widely used in computational fluid dynamics to ensure local conservation of mass, momentum, and energy. It discretizes the governing equations in integral form and solves them over discrete control volumes.

2 Coupling Mechanisms

In multiphase flow simulations, strong coupling between phases is necessary. Approaches such as synchronous and staggered coupling schemes are employed to iteratively solve the equations for shared interfaces.

3 Turbulence Models

Accurately capturing turbulence is crucial in multiphase flows, with models like k- and Large Eddy Simulation (LES) adapted to account for phase interactions. These models are integrated with multiphase frameworks to simulate real-world engineering problems.

Python Code Snippet

Below is a Python code snippet that encompasses the core computational elements needed to implement several key equations and algorithms mentioned in the chapter on multiphase flow quantification, including continuity, momentum, and energy equations, as well as different modeling strategies such as the Volume of Fluid (VOF) method and Eulerian models.

```python
import numpy as np

def continuity_equation(alpha_k, rho_k, u_k, Gamma_k, delta_t,
        delta_x):
    '''
    Calculate change in continuity for multiphase flow.
    :param alpha_k: Volume fraction of k-th phase.
    :param rho_k: Density of k-th phase.
    :param u_k: Velocity vector of k-th phase.
    :param Gamma_k: Inter-phase mass transfer.
    :param delta_t: Time step.
    :param delta_x: Spatial step.
    :return: Updated alpha_k * rho_k.
    '''
    divergence = np.gradient(alpha_k * rho_k * u_k, delta_x)
    return (alpha_k * rho_k + delta_t * (-divergence + Gamma_k))

def momentum_equation(alpha_k, rho_k, u_k, g, p, tau_k, F_kl,
        delta_t, delta_x):
    '''
    Calculate change in momentum for multiphase flow.
    :param alpha_k: Volume fraction of k-th phase.
    :param rho_k: Density of k-th phase.
    :param u_k: Velocity vector of k-th phase.
    :param g: Gravitational acceleration.
    :param p: Pressure.
    :param tau_k: Viscous stress tensor.
    :param F_kl: Inter-phase momentum exchange term.
    :param delta_t: Time step.
    :param delta_x: Spatial step.
    :return: Updated momentum.
    '''
    divergence = np.gradient(alpha_k * rho_k * u_k * u_k, delta_x)
    pressure_term = alpha_k * np.gradient(p, delta_x)
    viscous_term = np.gradient(tau_k, delta_x)
    return (alpha_k * rho_k * u_k + delta_t * (alpha_k * rho_k * g -
            pressure_term + viscous_term + F_kl - divergence))
```

```python
def energy_equation(alpha_k, rho_k, E_k, u_k, q_k, g, Phi_k, Q_kl,
    delta_t, delta_x):
    '''
    Calculate change in energy for multiphase flow.
    :param alpha_k: Volume fraction of k-th phase.
    :param rho_k: Density of k-th phase.
    :param E_k: Specific total energy.
    :param u_k: Velocity vector of k-th phase.
    :param q_k: Heat flux.
    :param g: Gravitational acceleration.
    :param Phi_k: Viscous dissipation.
    :param Q_kl: Inter-phase heat transfer.
    :param delta_t: Time step.
    :param delta_x: Spatial step.
    :return: Updated energy.
    '''
    divergence = np.gradient(alpha_k * rho_k * E_k * u_k, delta_x)
    heat_flux = np.gradient(q_k, delta_x)
    return (alpha_k * rho_k * E_k + delta_t * (-heat_flux + alpha_k
        * rho_k * u_k * g + Phi_k + Q_kl - divergence))

def vof_method(C, u, delta_t, delta_x):
    '''
    Apply the Volume of Fluid (VOF) method.
    :param C: Scalar function representing fluid phases.
    :param u: Velocity vector.
    :param delta_t: Time step.
    :param delta_x: Spatial step.
    :return: Updated scalar function C.
    '''
    return C - delta_t * np.gradient(u * C, delta_x)

def eulerian_eulerian_drag(C_D, rho_l, alpha_k, alpha_l, u_k, u_l,
    d_kl):
    '''
    Calculate drag force in Eulerian-Eulerian Models.
    :param C_D: Drag coefficient.
    :param rho_l: Liquid density.
    :param alpha_k: Volume fraction of k-th phase.
    :param alpha_l: Volume fraction of l-th phase.
    :param u_k: Velocity of k-th phase.
    :param u_l: Velocity of l-th phase.
    :param d_kl: Characteristic length scale.
    :return: Drag force vector.
    '''
    return (3/4) * C_D * rho_l * alpha_k * alpha_l * np.abs(u_k -
        u_l) * (u_k - u_l) / d_kl

def lagrangian_approach(m_p, F_p, delta_t):
    '''
    Calculate motion of particles using Lagrangian approach.
    :param m_p: Particle mass.
```

```
:param F_p: Forces acting on the particle.
:param delta_t: Time step.
:return: Updated particle velocity.
'''
u_p = np.zeros_like(F_p)
for i, F in enumerate(F_p):
    u_p[i] = u_p[i] + delta_t * F / m_p
return u_p

# Test case setup for demonstration
alpha_k = 0.6    # Volume fraction
rho_k = 1000     # Density in kg/m^3
u_k = np.array([1.0, 0.0, 0.0])   # Velocity vector in m/s
Gamma_k = 0.0    # No inter-phase mass transfer for simplicity
g = np.array([0.0, -9.81, 0.0])   # Gravity in m/s^2
p = 101325       # Pressure in Pa
tau_k = 0.0      # Viscous stress tensor simplification
F_kl = 0.0       # Inter-phase momentum exchange simplification
E_k = 5000       # Specific total energy in J/kg
q_k = 0.0        # Heat flux simplification
Phi_k = 0.0      # Viscous dissipation simplification
Q_kl = 0.0       # Inter-phase heat transfer simplification
C_D = 1.0        # Drag Coefficient
alpha_l = 0.4    # Other phase volume fraction
u_l = np.array([0.0, 0.0, 0.0])   # Other phase velocity vector
d_kl = 0.01      # Characteristic length scale in m
m_p = 1.0        # Particle mass in kg
F_p = np.array([1.0, 0.0, 0.0])   # Forces on particle in N
delta_t = 0.01   # Time step in seconds
delta_x = 0.01   # Spatial step in meters

# Example calculations
new_continuity = continuity_equation(alpha_k, rho_k, u_k, Gamma_k,
↪   delta_t, delta_x)
new_momentum = momentum_equation(alpha_k, rho_k, u_k, g, p, tau_k,
↪   F_kl, delta_t, delta_x)
new_energy = energy_equation(alpha_k, rho_k, E_k, u_k, q_k, g,
↪   Phi_k, Q_kl, delta_t, delta_x)
drag_force = eulerian_eulerian_drag(C_D, rho_k, alpha_k, alpha_l,
↪   u_k, u_l, d_kl)
particle_velocity = lagrangian_approach(m_p, F_p, delta_t)

print("Updated Continuity:", new_continuity)
print("Updated Momentum:", new_momentum)
print("Updated Energy:", new_energy)
print("Drag Force:", drag_force)
print("Particle Velocity:", particle_velocity)
```

This code includes key computational elements necessary for modeling multiphase flows:

- `continuity_equation` calculates changes in phase continuity using multiphase flow variables.

- `momentum_equation` computes momentum exchanges between phases considering forces and stresses.
- `energy_equation` determines phase energy updates incorporating heat, work, and inter-phase exchanges.
- `vof_method` represents the interface-tracking VOF method capturing immiscible fluid boundaries.
- `eulerian_eulerian_drag` models the drag force relying on phase interactions in Eulerian perspectives.
- `lagrangian_approach` simulates particle dynamics tracing individual particles or bubbles in fluid.

The code concludes with example outputs reflecting the application of these methods to a simplified test scenario.

Multiple Choice Questions

1. In the context of multiphase flow, what does the continuity equation primarily ensure?

    (a) Conservation of Momentum

    (b) Conservation of Mass

    (c) Conservation of Energy

    (d) Conservation of Volume

2. The momentum equation for multiphase flow involves which additional term to account for inter-phase interaction?

    (a) Pressure Gradient

    (b) Viscous Stress Tensor

    (c) Gravity

    (d) Inter-phase Momentum Exchange

3. What modeling approach in multiphase flow is particularly useful for tracking the interface between immiscible fluids?

    (a) Eulerian-Lagrangian Method

    (b) Eulerian-Eulerian Model

(c) Volume of Fluid (VOF) Method

(d) Discrete Phase Model

4. In an Eulerian-Eulerian approach, what role does the term $\mathbf{F}_{kl}^{drag}$ play?

    (a) It computes the heat exchange between phases

    (b) It calculates the mass transfer across phases

    (c) It models the drag force due to velocity differences between phases

    (d) It determines the total energy within a phase

5. Which numerical method is extensively used in simulating multiphase flow systems to ensure local conservation laws are satisfied?

    (a) Finite Element Method (FEM)

    (b) Boundary Element Method (BEM)

    (c) Finite Difference Method (FDM)

    (d) Finite Volume Method (FVM)

6. Which of the following describes a key characteristic of Lagrangian approaches to multiphase flow modeling?

    (a) Tracking each phase as a continuous field

    (b) Solving for volume fractions only

    (c) Managing particle-fluid interactions by tracking individual particles

    (d) Utilizing a single set of Navier-Stokes equations for all phases

7. In the context of multiphase flow, what does the scalar function C in the VOF method represent?

    (a) The concentration of the solid phase

    (b) The color coding of different phases

    (c) The volume fraction of a particular fluid phase

    (d) The energy density of the fluid mixture

Answers:
1. **B: Conservation of Mass**
The continuity equation governs the conservation of mass within a system. It ensures that the mass of each phase is conserved over time as it moves through space.
2. **D: Inter-phase Momentum Exchange**
The inter-phase momentum exchange term accounts for the transfer of momentum between different phases, which is crucial in capturing the dynamics involving interactions between phases.
3. **C: Volume of Fluid (VOF) Method**
The VOF method is designed for tracking and locating the free surface (or fluid-fluid interface) within immiscible multi-phase flows, which is suitable for problems where the position of the phase boundary is of interest.
4. **C: It models the drag force due to velocity differences between phases**
The drag term $\mathbf{F}_{kl}^{\text{drag}}$ is used to model the force exerted on one phase due to the relative motion with another phase, characterized by velocity differences between them.
5. **D: Finite Volume Method (FVM)**
FVM is a popular numerical method in computational fluid dynamics that focuses specifically on conserving quantities such as mass, momentum, and energy within control volumes, making it suitable for multiphase flow simulations.
6. **C: Managing particle-fluid interactions by tracking individual particles**
Lagrangian approaches focus on tracking the paths of individual particles, bubbles, or droplets (often in a fluid), which provides detailed information on the interactions with the surrounding fluid.
7. **C: The volume fraction of a particular fluid phase**
In the VOF method, the scalar function C represents the volume fraction of a given fluid phase within a computational cell, indicating how much of the cell is occupied by that phase.

Chapter 17

Compressible Flow – Isentropic Relations

Fundamentals of Compressible Flow

In the domain of fluid dynamics, compressible flows are characterized by significant changes in fluid density, distinguished from their incompressible counterparts. A pivotal aspect of these flows is the variable nature of speed of sound, which affects the Mach number—a dimensionless quantity defined as:

$$\text{Mach number } M = \frac{V}{a}$$

where V represents the flow velocity and a the local speed of sound. As flows transition through subsonic, transonic, supersonic, and hypersonic regimes, different physical phenomena occur, necessitating an understanding of isentropic processes.

Isentropic Flow Relations

Isentropic flows are idealized processes in which entropy remains constant. They provide valuable approximations for compressible flows in nozzles and diffusers. The isentropic relations are derived under the assumptions of adiabatic and reversible processes.

The pressure ratio in isentropic flows is given by:

$$\left(\frac{p}{p_0}\right) = \left(\frac{\rho}{\rho_0}\right)^\gamma = \left(\frac{T}{T_0}\right)^{\frac{\gamma}{\gamma-1}}$$

where p_0, ρ_0, and T_0 are the stagnation properties (pressure, density, and temperature), γ denotes the specific heat ratio, and p, ρ, and T are local fluid properties.

1 Temperature-Pressure Relationship

Another key isentropic relation is between temperature and pressure, which is critical in deriving velocity from thermodynamic state properties. This relation is expressed as:

$$T = T_0 \left(\frac{p}{p_0}\right)^{\frac{\gamma-1}{\gamma}}$$

This expression is fundamental when linking aerodynamic heating with pressure changes in compressible environments.

2 Density-Temperature Relationship

The density-temperature relationship under isentropic conditions serves as an important correlation for designing thermodynamic cycles and is stated as:

$$\rho = \rho_0 \left(\frac{T}{T_0}\right)^{\frac{1}{\gamma-1}}$$

This relation facilitates the calculation of mass flow rate in various flow regimes, which is crucial for engineering applications involving turbo-machinery or jet propulsion.

Application in Engineering Systems

In practice, isentropic flow equations offer analytical solutions that simplify the design and analysis of engineering systems. They are vital in:

1 Nozzle Design

For nozzles, the isentropic relations underpin the determination of exit velocities and cross-sectional area distributions necessary to achieve desired thrust levels. The velocity V at any point in a nozzle can be obtained using:

$$V = \sqrt{2c_p(T_0 - T)}$$

where c_p is the specific heat at constant pressure.

2 Diffuser Performance

In diffusers, isentropic equations help assess pressure recovery, which is crucial for the efficiency of jet engines. The static pressure recovery is evaluated to optimize performance while minimizing shock loss in supersonic diffusers.

3 Supersonic Flow Analysis

In supersonic flows, particularly around airfoils, the use of isentropic relations allows engineers to predict flow separation, shock wave formation, and associated changes in drag. The relationship between Mach number and area change governs the design of supersonic ducts:

$$\frac{A}{A_0} = \frac{1}{M}\left(\frac{2}{\gamma+1}\left(1 + \frac{\gamma-1}{2}M^2\right)\right)^{\frac{\gamma+1}{2(\gamma-1)}}$$

Where A is the cross-sectional area and A_0 is the reference area.

Isentropic Flow Realization and Limitations

While isentropic assumptions provide a simplified framework, real-world phenomena such as viscous effects, heat transfer, and shock waves necessitate the consideration of deviations from ideal isentropic behavior. Engineering methodologies incorporate empirical data and computational simulations to capture these complexities, ensuring designs remain robust under operational conditions.

Python Code Snippet

Below is a Python code snippet that encompasses core computational elements related to isentropic flow relations, including calculations for Mach number, pressure-temperature relationships, and applications for nozzle and diffuser design.

```python
import numpy as np

def mach_number(V, a):
    '''
    Calculate the Mach number.
    :param V: Flow velocity.
    :param a: Speed of sound.
    :return: Mach number.
    '''
    return V / a

def isentropic_pressure_ratio(p, p0, gamma):
    '''
    Calculate the isentropic pressure ratio.
    :param p: Local pressure.
    :param p0: Stagnation pressure.
    :param gamma: Specific heat ratio.
    :return: Isentropic pressure ratio.
    '''
    return (p / p0) ** (gamma / (gamma - 1))

def isentropic_temperature_relation(T0, p, p0, gamma):
    '''
    Calculate the temperature from isentropic relation.
    :param T0: Stagnation temperature.
    :param p: Local pressure.
    :param p0: Stagnation pressure.
    :param gamma: Specific heat ratio.
    :return: Local temperature.
    '''
    return T0 * (p / p0) ** ((gamma - 1) / gamma)

def isentropic_density_temperature_relation(rho0, T, T0, gamma):
    '''
    Calculate the density from isentropic relation.
    :param rho0: Stagnation density.
    :param T: Local temperature.
    :param T0: Stagnation temperature.
    :param gamma: Specific heat ratio.
    :return: Local density.
    '''
    return rho0 * (T / T0) ** (1 / (gamma - 1))

def nozzle_exit_velocity(T0, T, cp):
```

```python
    '''
    Calculate the exit velocity from a nozzle.
    :param T0: Stagnation temperature.
    :param T: Local temperature.
    :param cp: Specific heat at constant pressure.
    :return: Exit velocity.
    '''
    return np.sqrt(2 * cp * (T0 - T))

# Example parameters
V = 340     # Velocity in m/s
a = 343     # Speed of sound in m/s
p = 1e5     # Local pressure in Pa
p0 = 1.2e5  # Stagnation pressure in Pa
T0 = 300    # Stagnation temperature in Kelvin
gamma = 1.4 # Specific heat ratio for air
rho0 = 1.225 # Stagnation density in kg/m^3
cp = 1005   # Specific heat capacity at constant pressure in J/(kg·K)

# Calculate various parameters
M = mach_number(V, a)
pressure_ratio = isentropic_pressure_ratio(p, p0, gamma)
T = isentropic_temperature_relation(T0, p, p0, gamma)
density = isentropic_density_temperature_relation(rho0, T, T0,
    gamma)
exit_velocity = nozzle_exit_velocity(T0, T, cp)

# Outputs for demonstration
print("Mach Number:", M)
print("Isentropic Pressure Ratio:", pressure_ratio)
print("Temperature:", T)
print("Density:", density)
print("Nozzle Exit Velocity:", exit_velocity)
```

This code defines several key functions necessary for analyzing isentropic flows:

- `mach_number` computes the Mach number using flow velocity and local speed of sound.

- `isentropic_pressure_ratio` calculates the pressure ratio for isentropic flows.

- `isentropic_temperature_relation` derives the local temperature from the isentropic pressure relation.

- `isentropic_density_temperature_relation` provides the density calculation based on temperature and stagnation conditions.

- `nozzle_exit_velocity` computes the velocity at the nozzle exit, which is critical for thrust determination in nozzles.

The final block of code provides example calculations for these relations using typical parameters for air, aiding in the real-world application of isentropic analysis in engineering scenarios.

Multiple Choice Questions

1. How is the Mach number defined in compressible flow analysis?

    (a) $\frac{P}{\rho a}$

    (b) $\frac{V}{a}$

    (c) $\frac{a}{V}$

    (d) $\frac{\rho a}{V}$

2. Which of the following describes the main characteristic of isentropic flows?

    (a) Constant heat transfer

    (b) Constant entropy

    (c) Constant viscosity

    (d) Constant velocity

3. In the context of isentropic relations, the temperature-pressure relationship is expressed by which equation?

    (a) $T = T_0 \left(\frac{p}{p_0}\right)^{\frac{\gamma-1}{\gamma}}$

    (b) $T = T_0 \left(\frac{p}{p_0}\right)^{\gamma}$

    (c) $T = T_0 \left(\frac{V}{a}\right)^2$

    (d) $T = T_0 \left(\frac{\gamma+1}{2} M^2\right)$

4. What role does the specific heat ratio (γ) play in the isentropic flow relations?

    (a) It determines the flow area

    (b) It defines the relation between Mach number and temperature

(c) It influences the stagnation properties

   (d) It remains constant irrespective of flow conditions

5. Which of the following applications can be analyzed using isentropic flow equations?

   (a) Laminar boundary layers

   (b) Hydraulic systems

   (c) Nozzle designs for jet engines

   (d) Steady incompressible flows

6. Which equation is utilized to calculate velocity in a nozzle given isentropic flow conditions?

   (a) $V = \frac{P^2}{2c_p T}$

   (b) $V = \sqrt{2c_p(T_0 - T)}$

   (c) $V = c_p \frac{\Delta P}{\Delta \rho}$

   (d) $V = \frac{T}{T_0 \left(\frac{p}{p_0}\right)}$

7. In the equation $\frac{A}{A_0} = \frac{1}{M}\left(\frac{2}{\gamma+1}\left(1 + \frac{\gamma-1}{2}M^2\right)\right)^{\frac{\gamma+1}{2(\gamma-1)}}$, what does $\frac{A}{A_0}$ represent?

   (a) Flow velocity ratio

   (b) Pressure loss

   (c) Area ratio for flow expansion or compression

   (d) Density ratio

Answers:

1. **B:** $\frac{V}{a}$ The Mach number is defined as the ratio of the flow velocity (V) to the local speed of sound (a).

2. **B: Constant entropy** Isentropic flows are characterized by constant entropy, representing idealized adiabatic and reversible processes.

3. **A:** $T = T_0 \left(\frac{p}{p_0}\right)^{\frac{\gamma-1}{\gamma}}$ This equation relates temperature to pressure in an isentropic process, indicating how they change with respect to one another.

4. **C: It influences the stagnation properties** The specific heat ratio (γ) affects computations of stagnation properties like temperature, pressure, and density in isentropic flows.

5. **C: Nozzle designs for jet engines** Isentropic flow equations are particularly useful for analyzing the expansion and compression of gases in nozzles.

6. **B:** $V = \sqrt{2c_p(T_0 - T)}$ This equation is used to calculate the velocity of gases in a nozzle under isentropic conditions based on stagnation and local temperatures.

7. **C: Area ratio for flow expansion or compression** The expression $\frac{A}{A_0}$ represents the ratio of cross-sectional areas in a flow, critical for analyzing changes in flow velocity and properties through expansions or compressions.

Chapter 18

Hydraulic Jump Analysis

Introduction to Hydraulic Jumps in Open Channels

In open channel flow systems, hydraulic jumps represent sudden transitions from supercritical to subcritical flow. Characterized by rapid changes in flow depth and velocity, these phenomena manifest as a turbulent roller, dissipating energy substantially. Equations governing hydraulic jumps are crucial for understanding energy dissipation and flow regulation in hydraulic infrastructures.

Fundamental Equations of Hydraulic Jumps

The hydraulic jump is analyzed via the conservation of mass and momentum principles, expressed in terms of flow continuity and force balance.

1 Continuity Equation

The conservation of mass across a hydraulic jump is expressed by the continuity equation:

$$Q = A_1 v_1 = A_2 v_2$$

where Q denotes the flow rate, A_1 and A_2 are the cross-sectional areas, and v_1 and v_2 are the flow velocities before and after the jump, respectively.

2 Momentum Equation

Applying momentum conservation, the force balance for a hydraulic jump gives the classic formulation:

$$\rho v_1^2 y_1 + \frac{\rho g y_1^2}{2} = \rho v_2^2 y_2 + \frac{\rho g y_2^2}{2}$$

Here, ρ represents the fluid density, g is the acceleration due to gravity, and y_1 and y_2 are the flow depths upstream and downstream of the jump, respectively.

Energy Dissipation in Hydraulic Jumps

Hydraulic jumps act as energy dissipators within fluid systems. Energy loss, pivotal in designing hydraulic structures, is evaluated using:

1 Energy Loss Computation

The energy loss across a hydraulic jump, based on specific energy principles, is given by:

$$\Delta E = E_1 - E_2$$

Considering specific energies E_1 and E_2, the equation simplifies to:

$$\Delta E = \frac{(v_1^2 - v_2^2)}{2g}$$

2 Relation Between Upstream and Downstream Depths

The conjugate depth relationship, derived from the momentum equation, provides insights into energy dissipation:

$$y_2 = \frac{y_1}{2}\left(\sqrt{1 + 8\mathrm{Fr}_1^2} - 1\right)$$

where $Fr_1 = \frac{v_1}{\sqrt{gy_1}}$ is the initial Froude number. This relation underscores the dependency of y_2 on y_1 and the flow's criticality.

Applications and Implications in Engineering Systems

Hydraulic jump behavior is integral to the design of energy dissipators, water conveyance systems, and spillways. Understanding the jump location and magnitude assists engineers in predicting flow behavior and ensuring structural stability while preventing erosion and structural damage. The dimensionless characteristics of hydraulic jumps, such as the Froude number, are paramount in scaling laboratory results to full-scale engineering applications.

Utilization of these equations in hydraulic jump studies results in optimized designs for efficient energy dissipation in open channel systems, with significant impacts on energy management and structural integrity in hydraulic engineering projects.

Python Code Snippet

Below is a Python code snippet that encompasses the core computational elements relating to hydraulic jump analysis including continuity equation, momentum equation, energy loss computation, and conjugate depth relationships.

```
import numpy as np

def continuity_equation(Q, A1, v1, A2):
    '''
    Apply continuity equation for hydraulic jump analysis.
    :param Q: Flow rate [m^3/s].
    :param A1: Cross-sectional area before the jump [m^2].
    :param v1: Flow velocity before the jump [m/s].
    :param A2: Cross-sectional area after the jump [m^2].
    :return: Velocity after the jump [m/s].
    '''
    v2 = Q / A2
    return v2

def momentum_equation(rho, v1, y1, v2, y2, g):
    '''
    Apply momentum equation for hydraulic jump analysis.
    :param rho: Fluid density [kg/m^3].
    :param v1: Velocity before the jump [m/s].
```

```python
        :param y1: Depth before the jump [m].
        :param v2: Velocity after the jump [m/s].
        :param y2: Depth after the jump [m].
        :param g: Acceleration due to gravity [m/s^2].
        :return: Verifies force balance across the jump.
        '''
        term1 = rho * v1**2 * y1 + (rho * g * y1**2) / 2
        term2 = rho * v2**2 * y2 + (rho * g * y2**2) / 2
        return np.isclose(term1, term2, atol=1e-5)

def energy_loss(v1, v2, g):
    '''
    Compute energy dissipation across a hydraulic jump.
    :param v1: Velocity before the jump [m/s].
    :param v2: Velocity after the jump [m/s].
    :param g: Acceleration due to gravity [m/s^2].
    :return: Energy loss [m].
    '''
    delta_E = (v1**2 - v2**2) / (2 * g)
    return delta_E

def conjugate_depth(y1, Fr1):
    '''
    Compute the downstream conjugate depth using the Froude number.
    :param y1: Depth before the jump [m].
    :param Fr1: Froude number before the jump.
    :return: Conjugate depth after the jump [m].
    '''
    y2 = (y1 / 2) * (np.sqrt(1 + 8 * Fr1**2) - 1)
    return y2

# Constants
rho = 1000   # Fluid density [kg/m^3]
g = 9.81     # Acceleration due to gravity [m/s^2]

# Input Values
Q = 1.5   # Flow rate [m^3/s]
A1 = 0.5  # Cross-sectional area before jump [m^2]
v1 = Q / A1  # Velocity before the jump [m/s]
A2 = 0.8  # Cross-sectional area after jump [m^2]
v2 = continuity_equation(Q, A1, v1, A2)

y1 = 0.3  # Depth before the jump [m]
y2 = conjugate_depth(y1, v1 / np.sqrt(g * y1))

# Outputs
print("Velocity after Jump:", v2)
print("Momentum Equation Balanced:", momentum_equation(rho, v1, y1,
↪ v2, y2, g))
print("Energy Loss across Jump:", energy_loss(v1, v2, g))
print("Conjugate Depth After Jump:", y2)
```

This code defines several key functions necessary for hydraulic jump analysis:

- `continuity_equation` calculates the velocity after the jump using the continuity equation.
- `momentum_equation` verifies the force balance across a hydraulic jump using the momentum equation.
- `energy_loss` computes the energy dissipation across the hydraulic jump.
- `conjugate_depth` determines the downstream depth after the jump using the Froude number.

The final block of code provides examples of applying these elements to calculate various characteristics of a hydraulic jump, providing insight into velocity changes, energy loss, and conjugate depths.

Multiple Choice Questions

1. What physical phenomenon does a hydraulic jump in open channels represent?

    (a) Transition from subcritical to supercritical flow

    (b) Transition from supercritical to subcritical flow

    (c) Constant flow regime with depth fluctuations

    (d) Laminar flow transitioning into vortex shedding

2. Which equation expresses the principle of mass conservation in hydraulic jumps?

    (a) Bernoulli's equation

    (b) Continuity equation

    (c) Energy equation

    (d) Navier-Stokes equation

3. What does the initial Froude number (Fr_1) indicate in the context of hydraulic jumps?

    (a) The ratio of inertial forces to viscous forces

(b) The ratio of flow velocity to sound speed in the fluid

(c) The ratio of inertial forces to gravitational forces

(d) The ratio of potential energy to kinetic energy

4. Which principle is primarily used to derive the energy dissipation equation across hydraulic jumps?

   (a) Conservation of mass

   (b) Conservation of momentum

   (c) Conservation of energy

   (d) Conservation of angular momentum

5. The conjugate depth relationship in hydraulic jumps is primarily derived from which principle?

   (a) Bernoulli's principle

   (b) Energy conservation

   (c) Momentum conservation

   (d) No-slip condition

6. Why are hydraulic jumps important in engineering applications?

   (a) They enhance laminar flow conditions

   (b) They reduce fluid velocity to prevent cavitation

   (c) They significantly dissipate energy and stabilize flow

   (d) They increase the efficiency of fluid transport systems

7. In a hydraulic jump, how is the downstream depth (y_2) linked to the initial Froude number (Fr_1)?

   (a) y_2 is independent of Fr_1

   (b) y_2 decreases with increasing Fr_1

   (c) y_2 varies inversely with the square of Fr_1

   (d) y_2 increases with increasing Fr_1

Answers:

1. **B: Transition from supercritical to subcritical flow** Hydraulic jumps occur when a fluid transitions from supercritical (fast and shallow) to subcritical (slow and deep) flow, dissipating energy in the process.

2. **B: Continuity equation** The mass conservation principle for hydraulic jumps is expressed by the continuity equation, stating that the flow rate remains constant before and after the jump.

3. **C: The ratio of inertial forces to gravitational forces** The Froude number compares inertial forces to gravitational forces, indicating the flow regime's stability and propensity for sudden transitions like hydraulic jumps.

4. **C: Conservation of energy** The conservation of energy, particularly the specific energy principles, is employed to evaluate energy loss across hydraulic jumps.

5. **C: Momentum conservation** The conjugate depth relationship, which helps determine downstream flow depth, is derived using the conservation of momentum principle.

6. **C: They significantly dissipate energy and stabilize flow** Hydraulic jumps are utilized to dissipate excess energy, therefore preventing potential structural damage in hydraulic structures and flow instability.

7. **D: y_2 increases with increasing Fr_1** The downstream depth (y_2) in a hydraulic jump is strongly dependent on the initial Froude number (Fr_1); higher Fr_1 values typically result in a larger downstream depth.

Chapter 19

Hydrodynamic Stability Theory

Introduction to Hydrodynamic Stability

Hydrodynamic stability theory investigates the response of fluid interfaces and flows to perturbations. In fluid mechanics, stability analysis is crucial for understanding transitions from laminar flow to turbulence and for predicting the behavior of fluid interfaces. The focus is on linear stability analysis, which examines whether small disturbances grow or decay over time.

Governing Equations of Stability Analysis

1 Linearization of the Navier-Stokes Equations

The Navier-Stokes equations are linearized to facilitate stability analysis in the presence of small perturbations. For an incompressible fluid, the perturbed velocity field $\mathbf{u} = \mathbf{U} + \mathbf{u}'$, where $\mathbf{U}$ is the base flow and $\mathbf{u}'$ represents the perturbation, leads to the linearized momentum equations:

$$\frac{\partial \mathbf{u}'}{\partial t} + (\mathbf{U} \cdot \nabla)\mathbf{u}' + (\mathbf{u}' \cdot \nabla)\mathbf{U} = -\nabla p' + \nu \nabla^2 \mathbf{u}'$$

Here, p' is the perturbed pressure, and ν is the kinematic viscosity. The equation is accompanied by the continuity equation $\nabla \cdot \mathbf{u}' = 0$.

2 Normal Mode Analysis

Normal mode analysis assumes perturbations in the form of normal modes $\mathbf{u}' = \hat{\mathbf{u}} e^{i(\mathbf{k}\cdot\mathbf{x} - \omega t)}$, where $\mathbf{k}$ and ω are the wave vector and frequency, respectively. Substituting these into the linearized equations yields an eigenvalue problem of the form:

$$\mathcal{L}(\hat{\mathbf{u}}) = \omega \hat{\mathbf{u}}$$

Where $\mathcal{L}$ is a linear operator determined by the linearized Navier-Stokes equations which governs the stability characteristics of fluid systems.

Rayleigh's Stability Equation for Inviscid Flows

In the context of inviscid flows, Rayleigh's stability equation provides a criterion for the stability of shear flows. The primary equation is derived by assuming a parallel flow $U(y)$ in the streamwise direction:

$$(U - c)(\phi'' - k^2 \phi) - U'' \phi = 0$$

Here, $c = \omega/k$ is the wave speed and $\phi(y)$ is the perturbation stream function. This equation highlights the critical role of the inflection point criterion, given by Rayleigh's criterion $U''(y_i) = 0$, where instability is likely in the presence of an inflection point y_i.

Orr-Sommerfeld Equation for Viscous Flows

For viscous flows, the Orr-Sommerfeld equation extends Rayleigh's equation to account for the effects of viscosity. The linearized Navier-Stokes equations yield the Orr-Sommerfeld equation for perturbations $\phi(y)$ and ω:

$$i\alpha \left[(U - c)(\phi'' - \alpha^2 \phi) - U'' \phi \right] = \frac{1}{Re} \phi^{(4)} - 2\alpha^2 \frac{1}{Re} \phi'' + \alpha^4 \frac{1}{Re} \phi$$

Where $\alpha = k$ is the streamwise wavenumber and Re is the Reynolds number. This governs the stability characteristics of shear flows and offers insights into the transition to turbulence.

Applications in Fluid Interfaces

The stability of fluid interfaces is crucial in various engineering systems. Interfacial stability involves coupling of hydrodynamics and surface tension effects. The application of linear stability analysis to interfacial systems gives rise to the Kelvin-Helmholtz instability and Rayleigh-Taylor instability.

1 Kelvin-Helmholtz Instability

The Kelvin-Helmholtz instability involves two parallel fluid streams in relative motion. The dispersion relation for such systems can be derived as:

$$\omega^2 = gk\left(\frac{\rho_1 - \rho_2}{\rho_1 + \rho_2}\right) + \sigma k^3 \left(\frac{\rho_1 + \rho_2}{\rho_1 \rho_2}\right)$$

Where ρ_1 and ρ_2 are the fluid densities, and σ is the surface tension. This relation elucidates the role of density contrast and surface tension in interface stability.

2 Rayleigh-Taylor Instability

Rayleigh-Taylor instability occurs when a denser fluid overlies a less dense fluid in a gravitational field. The amplification rate of perturbations with wavenumber k is characterized by:

$$\gamma^2 = gk\left(\frac{\rho_1 - \rho_2}{\rho_1 + \rho_2}\right)$$

This highlights gravitational forces and density differences as the primary drivers for instability, often leading to complex flows such as mixing layers.

Complex Variable Theory in Stability Analysis

Hydrodynamic stability is advanced through complex variable theory, often leveraging conformal mapping techniques and the Riemann-

Hilbert problem for complex potential flows. This mathematical framework assists in solving potential flow stability problems and deriving analytical solutions for intricate instability phenomena.

Python Code Snippet

Below is a Python code snippet that encompasses the core computational elements of hydrodynamic stability theory including the linearization of the Navier-Stokes equations, normal mode analysis, Rayleigh's stability equation for inviscid flows, the Orr-Sommerfeld equation for viscous flows, and applications in fluid interfaces.

```
import numpy as np
import scipy.linalg as la
import matplotlib.pyplot as plt
from scipy.integrate import odeint

def linearize_ns(U, k, nu):
    '''
    Linearize Navier-Stokes equation perturbations.
    :param U: Base flow velocity.
    :param k: Wave number.
    :param nu: Kinematic viscosity.
    :return: Linear operator matrix A.
    '''
    # Define linear operator A
    A = np.array([[U, -k**2], [-1, nu*k**2]])
    return A

def normal_mode_analysis(U, k, omega):
    '''
    Perform normal mode analysis.
    :param U: Base flow velocity.
    :param k: Wave vector.
    :param omega: Frequency.
    :return: Eigenvalues and eigenvectors.
    '''
    A = linearize_ns(U, k, 0)
    eigenvalues, eigenvectors = la.eig(A)
    return eigenvalues, eigenvectors

def rayleigh_stability(U, k):
    '''
    Solve Rayleigh's stability equation for inviscid flows.
    :param U: Velocity profile function.
    :param k: Wave number.
    :return: Stability condition based on inflection points.
    '''
    y = np.linspace(0, 1, 100)
```

```python
    U_prime = np.gradient(U(y), y)
    inflection_points =
    ↪   np.where(np.diff(np.sign(np.gradient(U_prime, y))))[0]
    return inflection_points.size > 0

def orr_sommerfeld(U, k, Re):
    '''
    Solve the Orr-Sommerfeld equation for viscous flows.
    :param U: Velocity profile.
    :param k: Streamwise wave number.
    :param Re: Reynolds number.
    :return: Eigenvalues (growth rates).
    '''
    y = np.linspace(-1, 1, 100)
    U_prime = np.gradient(U(y), y)
    U_double_prime = np.gradient(U_prime, y)
    D2 = np.gradient(np.gradient(np.eye(len(y)), y), y)
    A = -Re * (np.diag(U - 1j*k*Re*D2) +
    ↪   1j*k*np.diag(U_double_prime))
    B = Re * np.eye(len(y)) - 1j*k*D2
    eigenvalues, _ = la.eig(A, B)
    return eigenvalues

def surface_stability(g, rho1, rho2, sigma, k):
    '''
    Calculate Kelvin-Helmholtz and Rayleigh-Taylor instabilities.
    :param g: Gravitational acceleration.
    :param rho1: Density of the first fluid.
    :param rho2: Density of the second fluid.
    :param sigma: Surface tension.
    :param k: Wave number.
    :return: Growth rates for KH and RT instabilities.
    '''
    kh_growth_rate = np.sqrt(g*k*(rho1 - rho2)/(rho1 + rho2) +
    ↪   sigma*k**3*(rho1 + rho2)/(rho1 * rho2))
    rt_growth_rate = np.sqrt(g*k*(rho1 - rho2)/(rho1 + rho2))
    return kh_growth_rate, rt_growth_rate

# Example of a simple linear velocity profile
def U(y):
    return 1-y**2

# Parameters
k = 1.0     # wave number
g = 9.81    # gravitational acceleration
rho1 = 1000 # density of fluid 1
rho2 = 850  # density of fluid 2
sigma = 0.07 # surface tension
Re = 1000   # Reynolds number

# Eigen analysis for normal mode
eigenvalues, eigenvectors = normal_mode_analysis(U, k, 0)
print('Normal Mode Eigenvalues:', eigenvalues)
```

```
# Rayleigh's stability check
is_stable = rayleigh_stability(U, k)
print('Rayleigh Stability:', 'Stable' if is_stable else 'Unstable')

# Orr-Sommerfeld results
os_eigenvalues = orr_sommerfeld(U, k, Re)
print('Orr-Sommerfeld Eigenvalues:', os_eigenvalues)

# Instability calculations
kh_growth_rate, rt_growth_rate = surface_stability(g, rho1, rho2,
    sigma, k)
print('Kelvin-Helmholtz Growth Rate:', kh_growth_rate)
print('Rayleigh-Taylor Growth Rate:', rt_growth_rate)
```

This code defines several key functions necessary for the implementation and analysis of hydrodynamic stability:

- `linearize_ns` linearizes the Navier-Stokes equations based on perturbations around a base flow.

- `normal_mode_analysis` performs eigenvalue analysis to understand the stability characteristics in normal modes.

- `rayleigh_stability` checks the stability of inviscid flows using Rayleigh's criterion based on the inflection points.

- `orr_sommerfeld` solves the Orr-Sommerfeld equation to determine the stability in viscous flows.

- `surface_stability` calculates the growth rates for fluid interface instabilities like Kelvin-Helmholtz and Rayleigh-Taylor.

The final block of code provides examples of computing these elements using example data which reflect commonly studied hydrodynamic and interfacial stability scenarios.

Multiple Choice Questions

1. Which equation describes the growth or decay of small disturbances in fluid systems in terms of normal modes?

    (a) Euler equations

    (b) Rayleigh's stability equation

    (c) Navier-Stokes equations

(d) Orr-Sommerfeld equation

2. What is the primary significance of the inflection point in Rayleigh's criterion for inviscid flow stability?

    (a) It indicates the onset of rotating flow structures.
    (b) It shows where viscosity effects dominate.
    (c) It marks a point where flow velocity reaches its maximum.
    (d) It suggests potential instability in shear flows.

3. In the context of hydrodynamic stability, what does the Orr-Sommerfeld equation account for that Rayleigh's equation does not?

    (a) Inviscid flow properties
    (b) Viscosity effects in shear flow
    (c) Compressibility of fluid
    (d) Multiphase flow conditions

4. What role does the Reynolds number play in the Orr-Sommerfeld equation?

    (a) It measures flow compressibility.
    (b) It represents the ratio of inertia to viscosity forces, affecting stability.
    (c) It quantifies surface tension effects.
    (d) It indicates the density contrast in multiphase flows.

5. Which instability occurs when a less dense fluid is accelerated into a denser fluid by gravitational forces?

    (a) Kelvin-Helmholtz instability
    (b) Rayleigh-Taylor instability
    (c) Taylor-Couette instability
    (d) Marangoni instability

6. In normal mode analysis for stability, what do the terms $\mathbf{k}$ and ω represent in the expression $\mathbf{u}' = \hat{\mathbf{u}} e^{i(\mathbf{k}\cdot\mathbf{x} - \omega t)}$?

    (a) Velocity and pressure

(b) Wave vector and frequency

(c) Wavelength and amplitude

(d) Strain rate and stress

7. Why is complex variable theory useful in hydrodynamic stability analysis?

   (a) It simplifies the Navier-Stokes equations to algebraic form.

   (b) It allows for exact solutions to be derived directly from potential flow equations via mathematical transformations.

   (c) It ensures continuity and differentiation manageability in flow fields.

   (d) It incorporates quantum effects into classical fluid dynamics problems.

Answers:

1. **B: Rayleigh's stability equation** Rayleigh's stability equation is used in the form of normal modes to describe how perturbations grow or decay in inviscid shear flows.

2. **D: It suggests potential instability in shear flows** In Rayleigh's criterion, the presence of an inflection point in the velocity profile of shear flow is indicative of potential instability.

3. **B: Viscosity effects in shear flow** The Orr-Sommerfeld equation extends Rayleigh's stability analysis by including the effects of viscosity, which is essential for analyzing viscous flows.

4. **B: It represents the ratio of inertia to viscosity forces, affecting stability** The Reynolds number in the Orr-Sommerfeld equation quantifies the relative importance of inertial and viscous forces, influencing flow stability characteristics.

5. **B: Rayleigh-Taylor instability** Rayleigh-Taylor instability occurs when a denser fluid is accelerated by gravity into a lighter fluid, leading to instability at the interface.

6. **B: Wave vector and frequency** In normal mode analysis, k represents the wave vector, and ω represents the frequency, crucial for analyzing oscillatory behavior of perturbations.

7. **B: It allows for exact solutions to be derived directly from potential flow equations via mathematical transformations** Complex variable theory offers a powerful analytical toolset for dealing with potential flow problems and deducing solutions concerning stability.

Chapter 20

Laminar Flow: Exact Solutions

Introduction to Laminar Flow Regimes

Laminar flow, characterized by smooth and orderly fluid motion, arises in fluid systems where inertial forces are dominated by viscous forces. This regime is typically described by a Reynolds number (Re) below a critical threshold. The study of laminar flow extends to solving the Navier-Stokes equations under assumptions that allow for simplifications, leading to exact solutions.

Governing Equations for Laminar Flow

1   Navier-Stokes Equations for Incompressible Flow

For isothermal, incompressible, and steady-state conditions, the Navier-Stokes equations for a Newtonian fluid can be expressed as:

$$\frac{\partial \mathbf{u}}{\partial t} + (\mathbf{u} \cdot \nabla)\mathbf{u} = -\frac{1}{\rho}\nabla p + \nu \nabla^2 \mathbf{u}$$

where $\mathbf{u}$ is the velocity field, ρ is the fluid density, p is the pressure field, and ν is the kinematic viscosity. For laminar flows, these equations are further simplified to capture the dominance of viscous forces over inertial forces.

2 Continuity Equation for Incompressible Flow

The continuity equation enforces mass conservation and is given by:

$$\nabla \cdot \mathbf{u} = 0$$

This equation is pivotal in simplifying the Navier-Stokes equations when deriving exact solutions for specific geometries.

Exact Solutions in Simple Geometries

1 Hagen-Poiseuille Flow in Circular Pipes

In cylindrical coordinates, the axisymmetric steady flow of an incompressible viscous fluid through a long, straight pipe leads to Hagen-Poiseuille flow. The velocity profile $u(r)$ for a pipe of radius R can be derived as:

$$u(r) = \frac{\Delta p}{4\nu L}(R^2 - r^2)$$

where Δp is the pressure drop along the length L of the pipe, and r is the radial distance from the centerline.

2 Couette Flow Between Parallel Plates

Consider the unidirectional flow between two infinite parallel plates, one of which moves at a constant velocity U. The steady-state velocity profile $u(y)$ is described by:

$$u(y) = \frac{U}{h}y$$

where h is the distance between the plates, and y is the perpendicular distance from the stationary plate.

Laminar Flow Over a Flat Plate: Blasius Solution

The Blasius boundary layer solution addresses the laminar flow of a fluid over a semi-infinite flat plate aligned with the flow direction. The boundary layer equations simplify to:

$$f''' + \frac{1}{2}ff'' = 0$$

where f is a dimensionless stream function defined by:

$$u = Uf'(\eta), \quad \eta = \frac{y}{\sqrt{\nu x/U}}$$

Here, U represents the free stream velocity, and η is a similarity variable.

Stokes' First Problem: Plate Sudden Motion

Known as the Rayleigh problem, this outlines the behavior when an infinite plate starts moving impulsively in its own plane. The transient development of the velocity profile $u(y,t)$ is solved using:

$$u(y,t) = U\,\mathrm{erfc}\left(\frac{y}{2\sqrt{\nu t}}\right)$$

where 'erfc' denotes the complementary error function, and t is time. This solution captures the diffusion of momentum from the plate into the stationary fluid.

Laminar Flow Through Channels: Pouseuille and Plane Couette Plates

1 Plane Poiseuille Flow

In a channel bounded by two stationary parallel plates, pressure-driven flow—plane Poiseuille flow—can be mathematically described by the velocity profile:

$$u(y) = \frac{\Delta p}{2\mu}\left(1 - \frac{y^2}{h^2}\right)$$

where μ is the dynamic viscosity, and Δp is the pressure gradient.

2 Superposed Couette and Poiseuille Flow

In a situation combining both Couette and Poiseuille flow between two parallel plates, the velocity distribution displays a linear plus quadratic combination:

$$u(y) = Ay + \frac{B}{2}(h^2 - y^2)$$

where parameters A and B are determined by the respective boundary velocities and pressure gradient.

Complex Geometries: Limitations and Extensions

Solving Navier-Stokes equations for laminar flow in complex geometries often requires numerical approaches or perturbative methods. However, exact solutions serve as critical benchmarks for validating computational fluid dynamics (CFD) models and understanding fundamental flow behavior.

Python Code Snippet

Below is a Python code snippet that encompasses the core computational elements of laminar flow regime analysis, including the calculation of velocity profiles for different flow scenarios, which are essential in solving exact laminar flow problems in fluid dynamics.

```
import numpy as np
from scipy.special import erfc

def hagen_poiseuille_flow(pressure_drop, radius, viscosity, length,
    radial_distance):
    '''
    Calculate the velocity profile for Hagen-Poiseuille flow in a
       pipe.
    :param pressure_drop: Pressure drop along the pipe.
    :param radius: Radius of the pipe.
    :param viscosity: Kinematic viscosity of the fluid.
    :param length: Length of the pipe.
    :param radial_distance: Radial distance from the pipe
       centerline.
    :return: Velocity at given radial distance.
    '''
```

```python
    return (pressure_drop / (4 * viscosity * length)) * (radius**2 -
    ↪ radial_distance**2)

def couette_flow(plate_velocity, distance_between_plates,
↪ distance_from_stationary_plate):
    '''
    Calculate the velocity profile for Couette flow between parallel
    ↪ plates.
    :param plate_velocity: Velocity of the moving plate.
    :param distance_between_plates: Distance between the plates.
    :param distance_from_stationary_plate: Distance from the
    ↪ stationary plate.
    :return: Velocity at given distance from the stationary plate.
    '''
    return (plate_velocity / distance_between_plates) *
    ↪ distance_from_stationary_plate

def blasius_solution(stream_velocity, kinematic_viscosity,
↪ distance_from_leading_edge, perpendicular_distance):
    '''
    Approximation to the Blasius solution for a boundary layer over
    ↪ a flat plate.
    :param stream_velocity: Free stream velocity.
    :param kinematic_viscosity: Kinematic viscosity of the fluid.
    :param distance_from_leading_edge: Distance from the leading
    ↪ edge of the plate.
    :param perpendicular_distance: Perpendicular distance to the
    ↪ plate surface.
    :return: Approximation of stream function velocity component at
    ↪ given distances.
    '''
    eta = perpendicular_distance / np.sqrt(kinematic_viscosity *
    ↪ distance_from_leading_edge / stream_velocity)
    return stream_velocity * (2 *
    ↪ erfc(eta))*np.sqrt(cinmatic_viscosity *
    ↪ distance_from_leading_edge / stream_velocity)

def stokes_first_problem(plate_velocity, viscosity, time,
↪ distance_from_plate):
    '''
    Solution to Stokes' first problem for plate sudden motion.
    :param plate_velocity: Velocity of the plate.
    :param viscosity: Kinematic viscosity of the fluid.
    :param time: Time after the plate starts moving.
    :param distance_from_plate: Distance from the moving plate.
    :return: Velocity at given distance and time.
    '''
    return plate_velocity * erfc(distance_from_plate / (2 *
    ↪ np.sqrt(viscosity * time)))

# Example usage with parameters
pipe_velocity = hagen_poiseuille_flow(500, 0.05, 1.0e-6, 1.0, 0.01)
couette_velocity = couette_flow(2.0, 0.1, 0.05)
```

```
blasius_approx = blasius_solution(1.0, 1.0e-6, 0.5, 0.01)
stokes_velocity = stokes_first_problem(5.0, 1.0e-6, 0.1, 0.01)

print("Hagen-Poiseuille Velocity:", pipe_velocity)
print("Couette Flow Velocity:", couette_velocity)
print("Blasius Approx. Velocity:", blasius_approx)
print("Stokes First Problem Velocity:", stokes_velocity)
```

This code defines several key functions necessary for the implementation of laminar flow profiles in fluid dynamics applications:

- `hagen_poiseuille_flow` function computes the velocity profile for laminar flow in a cylindrical pipe, emphasizing the pressure-driven characteristic of Hagen-Poiseuille flow.

- `couette_flow` calculates the velocity distribution between two parallel plates when one plate is in motion, illustrating shear-driven flow.

- `blasius_solution` provides a numerical approximation of the velocity profile in the boundary layer over a flat plate, relevant for aerodynamic applications.

- `stokes_first_problem` calculates the diffusion of momentum for impulsively started motion of a flat plate, demonstrating unsteady flow effects.

This computational toolset enables validation and analysis of laminar flow regimes in various geometrical and physical configurations.

Multiple Choice Questions

1. What characterizes laminar flow in fluid systems?

    (a) Low kinematic viscosity

    (b) High Reynolds number

    (c) Dominance of inertial forces

    (d) Smooth and orderly fluid motion

2. Which equation is primarily used for mass conservation in incompressible laminar flow?

    (a) Navier-Stokes Equation

(b) Bernoulli's Equation

(c) Continuity Equation

(d) Euler Equation

3. In Hagen-Poiseuille flow, what is required to maintain the flow of an incompressible viscous fluid through a pipe?

   (a) A constant temperature gradient

   (b) A moving boundary wall

   (c) A pressure drop along the pipe length

   (d) A high Mach number

4. What is a key assumption in deriving the Blasius solution for flow over a flat plate?

   (a) The flow is compressible

   (b) The flow is unsteady

   (c) The boundary layer is thin and the flow is steady

   (d) The plate is moving in the direction opposite to the flow

5. The velocity profile for plane Couette flow is:

   (a) Quadratic

   (b) Linear

   (c) Exponential

   (d) Logarithmic

6. Which mathematical method is often employed when exact solutions are not feasible for complex geometries?

   (a) Laplace transform

   (b) Numerical simulation

   (c) Perturbation methods

   (d) Dimensional analysis

7. Stokes' first problem is also known as:

   (a) Hagen-Poiseuille flow

   (b) Rayleigh problem

   (c) Boundary layer flow

(d) Plane Poiseuille flow

Answers:
1. **D: Smooth and orderly fluid motion** Laminar flow is characterized by a smooth and orderly pattern of motion, where layers of fluid slide past one another without mixing, unlike turbulent flow.

2. **C: Continuity Equation** The continuity equation is fundamental in enforcing mass conservation for incompressible flows, ensuring the fluid density remains constant throughout the flow.

3. **C: A pressure drop along the pipe length** Hagen-Poiseuille flow is characterized by flow through a pipe driven by a constant pressure gradient, which is necessary to overcome viscous forces and maintain flow.

4. **C: The boundary layer is thin and the flow is steady** The Blasius solution makes these assumptions to simplify the boundary layer equations for solving the flow over a flat plate.

5. **B: Linear** The velocity profile in plane Couette flow is linear due to the constant velocity difference between the two plates in the flow direction.

6. **B: Numerical simulation** Numerical simulation is frequently used when solving Navier-Stokes and related equations in complex geometries, where exact analytical solutions are not possible.

7. **B: Rayleigh problem** Stokes' first problem or the Rayleigh problem describes the flow development when an infinite plate suddenly starts moving in its plane, exploring viscous diffusion effects.

Chapter 21

Wave Motion Equations in Fluids

Foundation of Wave Motion in Fluid Dynamics

Wave motion in fluid systems plays a pivotal role in various engineering applications, from aerospace to civil engineering. Fluid waves can broadly be classified as linear or nonlinear, with linear waves following the principle of superposition and nonlinear waves exhibiting complex interactions.

Governing Equations for Linear Waves

1 Linear Wave Equation

The wave equation governing linear wave propagation in an inviscid, incompressible fluid can be expressed as:

$$\frac{\partial^2 \phi}{\partial t^2} = c^2 \nabla^2 \phi$$

where ϕ is the velocity potential, and c is the wave speed. This equation assumes small amplitude disturbances and potential flow conditions.

2 Dispersion Relations for Linear Waves

The dispersion relation characterizes the relationship between wave frequency ω and wave number k:

$$\omega = ck$$

This fundamental relation defines how wave speed c varies with wave frequency and wave number, crucial for studying wave behaviors in different mediums.

Analysis of Nonlinear Wave Phenomena

1 Korteweg-de Vries (KdV) Equation

The KdV equation captures the dynamics of shallow water solitary waves (solitons) and is given by:

$$\frac{\partial \eta}{\partial t} + c_0 \frac{\partial \eta}{\partial x} + \frac{3}{2} c_0 \frac{\eta}{h} \frac{\partial \eta}{\partial x} + \frac{1}{6} c_0 h^2 \frac{\partial^3 \eta}{\partial x^3} = 0$$

where η is the wave elevation, c_0 is the linear long wave speed, and h is the water depth. The equation illustrates both nonlinearity and dispersion effects.

2 Nonlinear Schrödinger Equation

For deep water gravity waves, the nonlinear Schrödinger equation models wave packet evolution:

$$i \frac{\partial A}{\partial t} + \alpha \frac{\partial^2 A}{\partial x^2} + \beta |A|^2 A = 0$$

where A is the complex wave envelope, α represents dispersion, and β characterizes nonlinearity. The equation highlights modulation and instability phenomena in wave trains.

Wave Propagation in Stratified Fluids

1 Internal Wave Dynamics

Internal waves, propagating at interfaces in stratified fluids, obey the following equation:

$$\frac{\partial^2 \phi}{\partial t^2} = N^2 \nabla^2 \phi$$

where N is the buoyancy frequency. This emphasizes internal wave behavior impacted by density stratifications.

2 Boussinesq Approximation

To model long internal waves in stratified fluids, the Boussinesq approximation is applied, assuming:

$$\frac{\partial^2 \phi}{\partial t^2} = g\frac{\partial \eta}{\partial z} + c^2 \nabla^2 \phi$$

where g is gravitational acceleration, and η is the displacement amplitude. This accounts for variations in density with minimal compressibility effects.

Numerical Approaches for Complex Wave Systems

1 Finite Difference Methods

Finite difference schemes approximate wave equations, enabling the analysis of wave propagation in complex geometries. Discretization techniques such as the `Forward Time Centered Space` (`FTCS`) and `Lax-Wendroff` methods provide explicit schemes for solving dynamic wave scenarios.

2 Spectral Methods

Spectral methods capitalize on the Fourier series to model wave interactions in fluid systems, leveraging orthogonality of trigonometric functions to achieve high accuracy in wave modeling without excessive computational costs.

Applications of Wave Motion Equations in Engineering

Wave motion equations help design coastal structures, predict climate impacts, and understand fluid-structure interactions in engineering applications. The theoretical insights and numerical techniques developed for linear and nonlinear waves guide practical solutions where analytical results are unattainable.

Python Code Snippet

Below is a Python code snippet that demonstrates the numerical algorithms for solving and analyzing wave motion equations discussed in this chapter, including functions for both linear and nonlinear wave phenomena, as well as numerical methods for simulation.

```
import numpy as np
import matplotlib.pyplot as plt

def linear_wave_equation(x, t, c):
    '''
    Solve the linear wave equation for velocity potential.
    :param x: Spatial domain.
    :param t: Time domain.
    :param c: Wave speed.
    :return: Velocity potential phi.
    '''
    phi = np.sin(c * x - t)
    return phi

def dispersion_relation(k, c):
    '''
    Calculate wave frequency using dispersion relation.
    :param k: Wave number.
    :param c: Wave speed.
    :return: Wave frequency omega.
    '''
    omega = c * k
    return omega

def kdv_equation(eta, c0, h, dx, dt, steps):
    '''
    Simulate the Korteweg-de Vries equation for solitons.
    :param eta: Initial wave elevation.
    :param c0: Linear long wave speed.
    :param h: Water depth.
    :param dx: Spatial step size.
    :param dt: Time step size.
    :param steps: Number of time steps.
    :return: Updated wave elevation.
    '''
    for _ in range(steps):
        eta = eta - dt * (c0 * np.gradient(eta, dx) +
```

```
                        (3/2) * (c0 * eta / h) * np.gradient(eta,
                        ↪   dx) +
                        (1/6) * c0 * h**2 *
                        ↪   np.gradient(np.gradient(np.gradient(eta,
                        ↪   dx), dx), dx))
        return eta

def schrodinger_eq_deep_water(A, alpha, beta, x, t):
    '''
    Solve the nonlinear Schrödinger equation for wave packets.
    :param A: Initial wave envelope.
    :param alpha: Dispersion coefficient.
    :param beta: Nonlinearity coefficient.
    :param x: Spatial domain.
    :param t: Time domain.
    :return: Updated wave envelope.
    '''
    for _ in range(len(t)):
        A = A + alpha * np.gradient(np.gradient(A, x), x) + beta *
        ↪   np.abs(A)**2 * A
    return A

def finite_difference_wave_1D(u0, c, dx, dt, steps):
    '''
    Simulate wave propagation via finite difference method.
    :param u0: Initial condition function.
    :param c: Wave speed.
    :param dx: Spatial step size.
    :param dt: Time step size.
    :param steps: Number of time steps.
    :return: Wave profile as a list.
    '''
    u = u0.copy()
    u_prev = u0.copy()
    u_n = u0.copy()
    cdtdx_squared = (c * dt / dx) ** 2

    for _ in range(steps):
        u_n[1:-1] = (2 * u[1:-1] - u_prev[1:-1] +
                    cdtdx_squared * (u[2:] - 2*u[1:-1] + u[:-2]))
        u_prev = u.copy()
        u = u_n.copy()

    return u

def plot_wave(x, u, title):
    '''
    Plot the wave results.
    :param x: Spatial domain.
    :param u: Wave profile.
    :param title: Plot title.
    '''
    plt.plot(x, u)
```

```
    plt.title(title)
    plt.xlabel('Position')
    plt.ylabel('Wave Amplitude')
    plt.grid(True)
    plt.show()

# Parameters and initial conditions
x = np.linspace(0, 10, 100)
t = np.linspace(0, 5, 200)
initial_wave = np.sin(x)

# Linear wave example
phi = linear_wave_equation(x, t[0], c=1)
plot_wave(x, phi, 'Linear Wave Equation')

# Nonlinear KdV wave example
eta_initial = np.ones_like(x)   # Initial wave elevation
eta_kdv = kdv_equation(eta_initial, c0=1, h=1, dx=0.1, dt=0.05,
↪    steps=100)
plot_wave(x, eta_kdv, 'KdV Soliton Wave')

# Finite difference wave propagation
wave_fd = finite_difference_wave_1D(initial_wave, c=1, dx=0.1,
↪    dt=0.01, steps=100)
plot_wave(x, wave_fd, 'Finite Difference Wave Propagation')
```

This code defines several key functions necessary for the analysis of wave motion in fluids:

- `linear_wave_equation` solves the linear wave equation to provide insight into velocity potential across a given domain.

- `dispersion_relation` calculates the wave frequency using the dispersion relation, relating wave speed and wave number.

- `kdv_equation` simulates shallow water wave dynamics using the Korteweg-de Vries equation for soliton behavior.

- `schrodinger_eq_deep_water` models wave packet evolution in deep water gravity waves using the nonlinear Schrödinger equation.

- `finite_difference_wave_1D` implements a finite difference method to simulate wave propagation in a one-dimensional domain.

- `plot_wave` is used to visualize the wave profiles generated by the computations above.

The example applications at the end demonstrate the use of these functions for modeling different wave phenomena discussed in the chapter.

Multiple Choice Questions

1. Which of the following describes the primary assumption for the linear wave equation in fluid dynamics?

    (a) High amplitude disturbances and turbulent flow

    (b) Viscous forces and compressible fluids

    (c) Small amplitude disturbances and potential flow

    (d) Nonlinear interactions and high velocity flow

2. What is the relationship between wave frequency (ω) and wave number (k) known as?

    (a) Doppler effect

    (b) Dispersion relation

    (c) Snell's law

    (d) Fourier transformation

3. The Korteweg-de Vries (KdV) equation is primarily used to model which type of wave phenomena?

    (a) Deep water gravity waves

    (b) Shallow water solitary waves

    (c) Internal waves in stratified fluids

    (d) Surface waves in open oceans

4. In the context of nonlinear wave equations, what is the primary role of β in the nonlinear Schrödinger equation?

    (a) It accounts for wave dispersion

    (b) It represents fluid density

    (c) It characterizes nonlinearity

    (d) It determines wave speed

5. Which approximation is commonly used for modeling long internal waves in stratified fluids?

(a) Bernoulli's principle

(b) Boussinesq approximation

(c) Laplace's equation

(d) Navier-Stokes approximation

6. Which numerical method is particularly known for leveraging the Fourier series to model wave interactions?

   (a) Finite element method

   (b) Finite volume method

   (c) Spectral methods

   (d) Finite difference method

7. In engineering applications, wave motion equations help in understanding:

   (a) The magnetic properties of materials

   (b) The thermal expansion of metals

   (c) The design of optical fibers

   (d) Fluid-structure interactions

Answers:

1. **C: Small amplitude disturbances and potential flow** The linear wave equation assumes small amplitude disturbances and potential flow conditions, making it valid for inviscid, incompressible fluid scenarios.

2. **B: Dispersion relation** The dispersion relation is fundamental to understanding how wave speed varies with wave frequency and wave number, crucial in wave propagation analysis.

3. **B: Shallow water solitary waves** The KdV equation models shallow water solitary waves, or solitons, capturing the nonlinearity and dispersion properties in such systems.

4. **C: It characterizes nonlinearity** In the nonlinear Schrödinger equation, β characterizes the nonlinearity of the wave system, impacting wave packet evolution.

5. **B: Boussinesq approximation** The Boussinesq approximation is applied to model long internal waves in stratified fluids, simplifying the dynamics by accounting for density variation with minimal compressibility effects.

6. **C: Spectral methods** Spectral methods utilize the Fourier series, enabling precise modeling of wave interactions due to the orthogonality of trigonometric functions.

7. **D: Fluid-structure interactions** Wave motion equations are crucial in applications involving fluid-structure interactions, aiding in the design and analysis of systems affected by wave dynamics, such as coastal structures.

Chapter 22

Pipe Networks Hydraulic Analysis

Introduction to Hydraulic Networks

Hydraulic networks are intricate systems designed to transport fluids through pipelines to various destinations. These systems are governed by principles of fluid mechanics that ensure efficient flow distribution and pressure management.

Fundamental Equations for Node Analysis

In fluid mechanics, nodes in a hydraulic network represent junction points where mass and energy conservation laws apply.

1 Continuity Equation at Nodes

The continuity equation ensures mass conservation at each node. For an incompressible fluid, it is expressed as:

$$\sum_{i=1}^{m} Q_i = 0$$

where m is the number of pipes connected to the node, and Q_i is the flow rate through the ith pipe.

2 Energy Equation in Loops

The energy equation is pivotal in loop analysis for closed network circuits, given by:

$$\sum_{j=1}^{n} h_{fj} + \sum_{j=1}^{n} \frac{p_j}{\rho g} + \sum_{j=1}^{n} z_j = 0$$

where h_{fj} represents the head loss in the jth pipe, p_j is the pressure, z_j is the elevation head, ρ is the fluid density, and g is the acceleration due to gravity.

Head Loss Calculations

Head loss within a pipe network is a critical factor affecting system efficiency and is typically characterized by frictional losses along the pipe length.

1 Darcy-Weisbach Equation

The Darcy-Weisbach equation offers a fundamental approach for calculating head loss due to friction:

$$h_f = f \frac{L}{D} \frac{V^2}{2g}$$

where h_f is the friction head loss, f is the Darcy friction factor, L is the length of the pipe, D is the diameter, and V is the velocity of the fluid.

2 Empirical Correlations for Friction Factor

The friction factor f is often determined using the Colebrook-White equation for turbulent flow:

$$\frac{1}{\sqrt{f}} = -2 \log_{10} \left(\frac{\epsilon/D}{3.7} + \frac{2.51}{Re\sqrt{f}} \right)$$

where ϵ is the pipe roughness coefficient, D is the diameter, and Re is the Reynolds number.

Optimization Techniques in Pipeline Systems

Efficient hydraulic systems are achieved through optimization techniques focusing on minimizing energy losses and operational costs.

1 Flow Distribution Optimization

Optimization of flow distribution within a network can be approached by minimizing the cumulative energy dissipation. This is achieved by varying flow rates and pipe diameters subject to system constraints.

2 Linear Programming for Network Optimization

Linear programming offers a structured method for optimizing pipeline networks. An objective function representing costs or energy losses is defined, subject to linear constraints such as:

$$\text{Minimize:} \quad C = \sum_{k=1}^{o} c_k Q_k$$

$$\text{Subject to:} \quad A\mathbf{x} = \mathbf{b}$$

where C is the total cost, c_k is the cost coefficient for the kth pipe, and A is a matrix representing the network constraints.

Numerical Simulation of Pipe Networks

Simulation of hydraulic models facilitates the evaluation of network performance under varied conditions and configurations.

1 Finite Element Analysis

Finite Element Analysis (FEA) is employed for detailed studies of flow distribution and pressure gradients within complex network geometries. Through discretization, it solves the governing equations iteratively.

2 Computational Fluid Dynamics (CFD) Approaches

CFD tools simulate intricate flow scenarios by solving the Navier-Stokes equations across the domain:

$$\frac{\partial \mathbf{u}}{\partial t} + \mathbf{u} \cdot \nabla \mathbf{u} = -\frac{1}{\rho}\nabla p + \nu \nabla^2 \mathbf{u}$$

where $\mathbf{u}$ is the velocity vector, ν is the kinematic viscosity, and p is the pressure.

Python Code Snippet

Below is a Python code snippet that encompasses the core computational elements related to hydraulic networks, including continuity equation checks, head loss calculations, flow optimization, and numerical simulation with both Finite Element Analysis and Computational Fluid Dynamics approaches.

```python
import numpy as np
from scipy.optimize import linprog
from scipy import sparse
from scipy.sparse.linalg import spsolve

def continuity_at_nodes(flow_rates):
    '''
    Check mass conservation at network nodes using the continuity
        equation.
    :param flow_rates: List of flow rates for each pipe in the
        network.
    :return: True if mass conservation holds at all nodes, False
        otherwise.
    '''
    return np.isclose(np.sum(flow_rates), 0)

def darcy_weisbach_head_loss(f, L, D, V):
    '''
    Calculate head loss using Darcy-Weisbach equation.
    :param f: Darcy friction factor.
    :param L: Length of the pipe.
    :param D: Diameter of the pipe.
    :param V: Velocity of the fluid.
    :return: Friction head loss.
    '''
    g = 9.81
    return f * (L / D) * (V**2) / (2 * g)

def pipe_flow_optimization(cost_coeffs, A, b):
```

```python
    '''
    Perform linear programming to optimize flow distribution in
    ↪    pipeline networks.
    :param cost_coeffs: Coefficient list representing cost for each
    ↪    pipe.
    :param A: Constraint matrix.
    :param b: Constraint boundary.
    :return: Optimal flow distribution across the network.
    '''
    res = linprog(cost_coeffs, A_eq=A, b_eq=b, bounds=(0, None))
    return res.x

def finite_element_analysis(nodes, elements, boundary_conditions):
    '''
    Utilize Finite Element Analysis to solve flow distribution in
    ↪    the network.
    :param nodes: List of nodes in the network.
    :param elements: List of element connectivity.
    :param boundary_conditions: Boundary conditions applied to the
    ↪    network.
    :return: Pressure distribution across the network.
    '''
    K = sparse.eye(len(nodes))    # Placeholder for assembled
    ↪    stiffness matrix
    F = np.zeros(len(nodes))    # Placeholder for external force
    ↪    vector
    F[boundary_conditions] = 1    # Example boundary conditions
    return spsolve(K, F)

def cfd_simulation(domain, boundary_conditions, U_inlet):
    '''
    Simulate complex fluid flow using CFD by solving Navier-Stokes
    ↪    equations.
    :param domain: Definition of the simulation domain.
    :param boundary_conditions: Boundary conditions setup.
    :param U_inlet: Inlet velocity conditions.
    :return: Velocity field, Pressure field.
    '''
    # Placeholder for CFD meshgrid and solution (e.g., using
    ↪    OpenFOAM)
    velocity_field = np.zeros(domain)
    pressure_field = np.zeros(domain)
    # Pseudo-implementation of the CFD processing step
    for _ in range(100):    # Example iteration for convergence
        velocity_field += U_inlet / 100    # Mock update
    return velocity_field, pressure_field

# Example application
flow_rates_example = [3, -1, -2]    # Arbitrary example flow rates
L, D, V, f = 1000, 0.5, 4, 0.02
cost_coeffs, A, b = [1, 2, 3], [[1, 0, 1], [0, 1, -1]], [0, 0]
nodes_example = [0, 1, 2]
elements_example = [(0, 1), (1, 2)]
```

```
boundary_conditions_example = [0, 2]
domain_example = (10, 10)

mass_conservation = continuity_at_nodes(flow_rates_example)
head_loss = darcy_weisbach_head_loss(f, L, D, V)
optimal_flow = pipe_flow_optimization(cost_coeffs, A, b)
pressure_dist = finite_element_analysis(nodes_example,
↪   elements_example, boundary_conditions_example)
velocity_field, pressure_field = cfd_simulation(domain_example,
↪   boundary_conditions_example, 1.5)

print("Mass Conservation:", mass_conservation)
print("Head Loss:", head_loss)
print("Optimal Flow Distribution:", optimal_flow)
print("Pressure Distribution:", pressure_dist)
print("Velocity Field:", velocity_field)
print("Pressure Field:", pressure_field)
```

This code defines several key functions necessary for analyzing and optimizing hydraulic networks:

- `continuity_at_nodes` function checks the mass conservation at nodes based on flow rates.

- `darcy_weisbach_head_loss` calculates head loss due to friction using the Darcy-Weisbach equation.

- `pipe_flow_optimization` employs linear programming for optimizing flow distribution under specified constraints.

- `finite_element_analysis` sets up a basic example for using Finite Element Analysis to solve flow problems in networks.

- `cfd_simulation` provides a placeholder for complex Computational Fluid Dynamics simulation of flow fields in a defined domain.

The final block demonstrates the use of these functions with sample data to calculate flow characteristics and solve optimization problems in hydraulic systems.

Multiple Choice Questions

1. In fluid mechanics, nodes within a hydraulic network are significant because they:

    (a) Determine the overall diameter of piping in the system.

(b) Represent junction points where mass and energy conservation laws apply.

(c) Serve as exclusive entry points for external fluid sources.

(d) Indicate areas with the highest pressure losses.

2. The continuity equation for incompressible fluids at a node is essential for ensuring:

    (a) Momentum conservation.

    (b) Energy conservation.

    (c) Mass conservation.

    (d) Viscosity conservation.

3. Which equation is pivotal when analyzing energy balance around a loop in a closed hydraulic network circuit?

    (a) Bernoulli's Equation.

    (b) Navier-Stokes Equation.

    (c) Darcy-Weisbach Equation.

    (d) Energy Equation in Loops.

4. For calculating the friction factor in turbulent flows within pipes, which of the following equations is commonly used?

    (a) Manning's Equation.

    (b) Hagen-Poiseuille Equation.

    (c) Colebrook-White Equation.

    (d) Bernoulli's Equation.

5. In the context of optimizing pipeline systems, linear programming is employed primarily to:

    (a) Maximize fluid velocity.

    (b) Minimize cumulative energy dissipation.

    (c) Calculate exact pipe diameters.

    (d) Ensure continuity of flow at each node.

6. Finite Element Analysis (FEA) is utilized in pipe network simulations mainly for:

    (a) Applying linear programming techniques.

- (b) Exact solution of Navier-Stokes equations.
- (c) Detailed studies of flow distribution and pressure gradients.
- (d) Direct measurement of pipe diameter and length.

7. The Darcy-Weisbach equation is used to calculate:
   - (a) The velocity profile of a fluid in a pipe.
   - (b) The head loss due to friction in a pipe.
   - (c) The pressure difference across a node.
   - (d) The energy loss due to sudden expansions in piping.

Answers:

1. **B: Represent junction points where mass and energy conservation laws apply.** Nodes are critical points in a hydraulic network where the principles of mass and energy conservation are enforced to ensure proper network operation.

2. **C: Mass conservation.** The continuity equation ensures that the mass entering and leaving a node is balanced, maintaining mass conservation across the network.

3. **D: Energy Equation in Loops.** The energy equation for loops ensures that the sum of head losses, pressures, and elevation heads within a closed loop balances, facilitating energy conservation in the network.

4. **C: Colebrook-White Equation.** The Colebrook-White equation is widely used for calculating the friction factor in turbulent pipe flow conditions, accounting for pipe roughness.

5. **B: Minimize cumulative energy dissipation.** Linear programming in hydraulic network optimization focuses on reducing energy losses, thereby improving the system's efficiency and reducing operational costs.

6. **C: Detailed studies of flow distribution and pressure gradients.** FEA helps analyze the distribution of fluid flow and evaluate pressure gradients within complicated network geometries.

7. **B: The head loss due to friction in a pipe.** The Darcy-Weisbach equation provides a way to calculate the frictional head loss, critical for assessing the energy efficiency of fluid transport in the network.

Chapter 23

Stream Function and Complex Potential

Two-Dimensional Flow Analysis

The study of two-dimensional flow problems in fluid mechanics can be simplified using stream function dynamics and complex potential theory. These tools provide valuable insights into analyzing and solving complex flow patterns.

1 Stream Function Dynamics

The stream function, denoted as ψ, is a mathematical construct used to represent two-dimensional, incompressible flow fields. It is particularly useful because the continuity equation is inherently satisfied. For a two-dimensional flow with velocity components u and v, the stream function is defined by:

$$u = \frac{\partial \psi}{\partial y}$$

$$v = -\frac{\partial \psi}{\partial x}$$

By these definitions, the continuity equation for incompressible flow, $\frac{\partial u}{\partial x} + \frac{\partial v}{\partial y} = 0$, is automatically satisfied. Isolines of ψ represent streamlines, thus providing a direct method of visualizing the flow pattern.

The utility of the stream function extends into the Laplace domain, where it becomes crucial in solving Laplace's equation $\nabla^2 \psi = 0$, indicative of irrotational and incompressible flows.

2 Complex Potential Applications

Complex potential theory is another powerful analytical method used in addressing two-dimensional irrotational flow problems. In complex analysis, the potential function Φ is defined as:

$$\Phi = \phi + i\psi$$

where ϕ is the velocity potential satisfying $\nabla^2 \phi = 0$, and ψ is the stream function. The physical significance of Φ lies in its ability to encapsulate both potential flow and stream function dynamics, with the real part representing velocity potential lines and the imaginary part representing streamlines.

In the context of complex analysis, the complex potential is exploited through the complex conjugate $\bar{W}(z) = \frac{d\Phi}{dz}$, where $z = x + iy$. This notation is pivotal in deriving the velocity field:

$$\frac{d\Phi}{dz} = u - iv$$

This makes the analysis elegant in that conformal mappings transform complex potential functions from one domain to another, preserving angles and offering solutions to boundary value problems in complex geometries.

Conformal Mapping Techniques

Conformal mapping plays a crucial role in transforming complex flow domains into simpler geometries, enabling analytical solutions for flow over irregular objects such as airfoils or through curved channels.

1 Mathematical Formulation of Conformal Maps

Conformal mapping transforms complex variables in a manner that preserves the angular relationship at each point. Mathematically, a conformal map from domain z to domain ζ is expressed by:

$$\zeta = f(z)$$

where $f(z)$ is an analytic function and mapping is invertible. For flow analysis, selected mappings such as the Joukowski transformation, defined by:

$$\zeta = z + \frac{1}{z}$$

translate circular domains into airfoil shapes, while maintaining analytical tractability.

2 Applications in Engineering Problems

In engineering applications, conformal mappings are instrumental in solving potential flow problems around airfoils, cylinders, and within ducts. By transforming boundaries to simpler geometric configurations, one can directly apply known solutions and subsequently invert the mapping for results in the original domain.

Practical Implications of Stream Function and Complex Analysis

The implementation of stream function and complex potential in solving practical fluid dynamics problems spans across a variety of engineering fields, including aerodynamics, hydraulics, and computational fluid dynamics (CFD).

1 Flow Around Bodies and Obstacles

Stream function formulations allow for the direct computation of flow patterns around submerged bodies and obstacles. By employing boundary conditions along the bodies' surfaces, engineers utilize ψ to predict circulation and pressure distributions, essential for optimizing design in aerospace and marine engineering.

2 Integration into Computational Methods

In CFD, the incorporation of the stream function and complex potential into numerical algorithms enhances the stability and accuracy of solvers applied to potential flow regions. For instance, in boundary element methods (BEM), complex potential enables boundary representations that simplify the solution of otherwise cumbersome velocity fields.

The meticulous application of these mathematical tools leads to innovative solutions and robust designs in engineering practice.

Python Code Snippet

Below is a Python code snippet that illustrates the core computational elements for analyzing two-dimensional flow using stream function and complex potential methods, including the mathematical formulation of stream functions, complex potential, and conformal mapping techniques.

```python
import numpy as np
import matplotlib.pyplot as plt

def stream_function(x, y):
    '''
    Calculate the stream function for a two-dimensional
      incompressible flow.
    :param x: X-coordinate in the flow field.
    :param y: Y-coordinate in the flow field.
    :return: Stream function value.
    '''
    return y*x  # Example function

def velocity_field(x, y):
    '''
    Derive velocity components from the stream function.
    :param x: X-coordinate in the flow field.
    :param y: Y-coordinate in the flow field.
    :return: Tuple of velocity components (u, v).
    '''
    psi = stream_function(x, y)
    u = np.gradient(psi, axis=0)   # approximate /y
    v = -np.gradient(psi, axis=1)  # approximate -/x
    return u, v

def complex_potential(z):
    '''
    Define a complex potential for the flow.
    :param z: Complex variable (x + iy).
    :return: Complex potential value.
    '''
    return z + 1/z  # Example corresponding to a doublet

def conformal_map(z):
    '''
    Apply a conformal map to transform the flow field.
    :param z: Complex variable (x + iy).
    :return: Transformed complex variable.
```

```python
    '''
    return z + 1/z

def plot_streamlines(x, y, u, v):
    '''
    Plot the streamlines of the flow.
    :param x: Grid of x-coordinates.
    :param y: Grid of y-coordinates.
    :param u: X-component of velocity.
    :param v: Y-component of velocity.
    '''
    plt.figure(figsize=(8, 6))
    plt.streamplot(x, y, u, v)
    plt.title('Streamlines')
    plt.xlabel('x')
    plt.ylabel('y')
    plt.grid(True)
    plt.show()

# Example usage
x = np.linspace(-2, 2, 100)
y = np.linspace(-2, 2, 100)
X, Y = np.meshgrid(x, y)
# Compute velocity field
U, V = velocity_field(X, Y)
# Plotting
plot_streamlines(x, y, U, V)
```

This code demonstrates the fundamental techniques for implementing stream function and complex potential to analyze two-dimensional flows:

- The `stream_function` function defines a basic stream function for incompressible flow.

- The `velocity_field` function computes velocity components (u, v) from the stream function.

- The `complex_potential` function sets up a complex potential for flow representation using complex variables.

- The `conformal_map` function applies a conformal mapping, useful in transforming complex geometries for different flow fields.

- The `plot_streamlines` function visualizes the flow by plotting streamlines on a defined grid.

The provided example utilizes dummy data for demonstrating the calculation and visualization of streamlines within a flow field, highlighting the importance of these computational techniques in fluid dynamics analysis.

Multiple Choice Questions

1. Which of the following is true regarding the stream function ψ in two-dimensional incompressible flow?

    (a) It is defined for compressible flow fields.

    (b) Its isolines represent equipotential lines.

    (c) It satisfies the continuity equation inherently.

    (d) It is only applicable to three-dimensional flows.

2. In complex potential theory, the potential function Φ is expressed as:

    (a) $\Phi = \phi \cdot \psi$

    (b) $\Phi = \phi - i\psi$

    (c) $\Phi = \psi + i\phi$

    (d) $\Phi = \phi + i\psi$

3. The main advantage of using conformal mapping in fluid mechanics is to:

    (a) Calculate three-dimensional fluid flows.

    (b) Solve potential flow problems in complex geometries.

    (c) Determine temperature distributions in fluids.

    (d) Analyze compressible shock waves.

4. Which equation does the stream function ψ satisfy in irrotational flow?

    (a) The Navier-Stokes equation.

    (b) The Euler equation.

    (c) The Laplace equation.

    (d) The Bernoulli equation.

5. What is the mathematical expression for the Joukowski transformation?

    (a) $\zeta = z + z^2$

(b) $\zeta = z^2 - z$
(c) $\zeta = z - \frac{1}{z}$
(d) $\zeta = z + \frac{1}{z}$

6. In two-dimensional flow analysis, the velocity components u and v in terms of the stream function ψ are given by:

    (a) $u = \frac{\partial \psi}{\partial x}, v = \frac{\partial \psi}{\partial y}$
    (b) $u = \frac{\partial \psi}{\partial y}, v = -\frac{\partial \psi}{\partial x}$
    (c) $u = -\frac{\partial \psi}{\partial y}, v = \frac{\partial \psi}{\partial x}$
    (d) $u = \frac{\partial^2 \psi}{\partial x \partial y}, v = \frac{\partial^2 \psi}{\partial y \partial x}$

7. The complex conjugate of the complex potential $\bar{W}(z)$ used in fluid flow analysis represents which of the following?

    (a) Pressure distribution in the flow.
    (b) Velocity field in the flow.
    (c) Streamlines in a compressible flow.
    (d) Temperature distribution across the flow.

Answers:

1. **C: It satisfies the continuity equation inherently** The stream function ψ inherently satisfies the continuity equation for two-dimensional incompressible flows by its definition.

2. **D: $\Phi = \phi + i\psi$** The complex potential function Φ is given by this expression, where ϕ is the velocity potential and ψ is the stream function.

3. **B: Solve potential flow problems in complex geometries** Conformal mapping is used to simplify complex domains into simpler ones, making it easier to solve potential flow problems analytically.

4. **C: The Laplace equation** In irrotational flow, the stream function ψ satisfies the Laplace equation, indicating that both ψ and the velocity potential are harmonic functions.

5. **D: $\zeta = z + \frac{1}{z}$** This is the mathematical expression for the Joukowski transformation, valuable in transforming circular shapes into airfoils.

6. **B: $u = \frac{\partial \psi}{\partial y}, v = -\frac{\partial \psi}{\partial x}$** These are the velocity components u and v in terms of the stream function ψ, which help in visualizing flow patterns.

7. B: Velocity field in the flow The derivative of the complex potential function $\frac{d\Phi}{dz}$ represents the velocity field, expressed in terms of complex variables.

Chapter 24

Flow-Induced Vibrations

Introduction to Flow-Induced Vibrations

Flow-induced vibrations (FIV) arise due to the complex interaction between fluid flow and structural dynamics. Understanding these phenomena requires a multidisciplinary approach, encompassing fluid dynamics, structural mechanics, and vibration analysis. Such vibrations can lead to significant engineering challenges, including fatigue failure, noise generation, and system instability.

Governing Equations

The behavior of flow-induced vibrations is dictated by a set of coupled equations involving fluid and structure interactions. The Navier-Stokes equations govern the fluid dynamics, expressed in their incompressible form as:

$$\frac{\partial \mathbf{u}}{\partial t} + \mathbf{u} \cdot \nabla \mathbf{u} = -\frac{1}{\rho}\nabla p + \nu \nabla^2 \mathbf{u} \qquad (24.1)$$

where $\mathbf{u}$ represents the velocity vector, p is the pressure, ρ is the fluid density, and ν is the kinematic viscosity.

The structural response to fluid forces can be modeled using equations of motion for elastic structures. For a linear elastic structure, the equation of motion is given by:

$$m\frac{d^2\mathbf{x}}{dt^2} + c\frac{d\mathbf{x}}{dt} + k\mathbf{x} = \mathbf{f}_{\text{fluid}} \qquad (24.2)$$

where m, c, and k denote the mass, damping, and stiffness matrices of the structure, respectively, and $\mathbf{f}_{\text{fluid}}$ represents the fluid forces acting on the structure.

Fluid-Structure Interaction Mechanisms

The coupling between fluid dynamics and structural response is central to the analysis of FIV. The interaction can be characterized by mechanisms such as vortex shedding, buffeting, and flutter.

1 Vortex-Induced Vibrations

Vortex-induced vibrations (VIV) often occur when the fluid flow past a bluff body induces the periodic shedding of vortices. The frequency of vortex shedding, known as the Strouhal frequency f_s, is described by the Strouhal number relation:

$$St = \frac{f_s D}{U} \qquad (24.3)$$

where D is the characteristic length (such as diameter of a cylinder), and U is the free-stream velocity.

The lift force on the body fluctuates sinusoidally as:

$$F_L(t) = \frac{1}{2}\rho U^2 A C_L \sin(2\pi f_s t) \qquad (24.4)$$

where A is the reference area, and C_L is the lift coefficient.

2 Flutter Phenomena

Flutter is an aeroelastic instability that occurs when aerodynamic forces couple with the natural vibration modes of a structure, leading to divergent oscillations. The critical flutter speed U_{cr} is a key parameter and is determined by analyzing the system's dynamic stability:

$$\det(\mathbf{K}_{\text{eff}} - \omega^2 \mathbf{M}_{\text{eff}} + i\omega \mathbf{C}_{\text{eff}}) = 0 \qquad (24.5)$$

where $\mathbf{K}_{\text{eff}}$, $\mathbf{M}_{\text{eff}}$, and $\mathbf{C}_{\text{eff}}$ are the effective stiffness, mass, and damping matrices that incorporate fluid-structure interaction effects.

3 Buffeting and Unsteady Aerodynamics

Buffeting involves random, turbulent fluctuations in aerodynamic forces due to unsteady flow separation. This is often analyzed using unsteady aerodynamic models, such as the Wagner function for lift response:

$$C_L(t) = \int_0^t \dot{\phi}(t-\tau) V_\infty^2(\tau) \, d\tau \qquad (24.6)$$

where $\phi(t)$ is the wagner function that describes lift response over time, and $V_\infty(\tau)$ is the instantaneous free-stream velocity.

Impact on Structural Design

The occurrence of flow-induced vibrations necessitates careful attention to the design of structures to mitigate detrimental effects. Considerations in engineering design include tuning the natural frequencies of structures to avoid resonance conditions, optimizing structural shape to minimize adverse flow phenomena, and implementing damping mechanisms to absorb vibrational energy.

The integration of fluid-structure interaction analysis into the design and evaluation stages helps ensure safety, efficiency, and longevity in applications ranging from bridge design to aerospace structures. Advanced numerical methods, such as the finite element method for structures along with computational fluid dynamics (CFD), play a crucial role in the accurate prediction and analysis of flow-induced vibration effects.

Python Code Snippet

Below is a Python code snippet that encompasses the core computational elements for analyzing flow-induced vibrations, including calculation of vortex shedding frequency, lift force, and evaluation of flutter conditions.

```
import numpy as np
from scipy.linalg import eig

def strouhal_frequency(free_stream_velocity, diameter,
    strouhal_number=0.2):
    '''
    Calculate the Strouhal frequency for vortex-induced vibration.
```

```
    :param free_stream_velocity: Free-stream velocity of the fluid.
    :param diameter: Characteristic length.
    :param strouhal_number: Strouhal number.
    :return: Frequency of vortex shedding.
    '''
    return strouhal_number * free_stream_velocity / diameter

def lift_force_fluctuation(density, free_stream_velocity, area,
↪ lift_coefficient, frequency, time):
    '''
    Calculate the fluctuating lift force on a body.
    :param density: Fluid density.
    :param free_stream_velocity: Free-stream velocity.
    :param area: Reference area.
    :param lift_coefficient: Lift coefficient.
    :param frequency: Frequency of lift force oscillation.
    :param time: Time variable.
    :return: Lift force at given time.
    '''
    return 0.5 * density * free_stream_velocity**2 * area *
↪ lift_coefficient * np.sin(2 * np.pi * frequency * time)

def flutter_analysis(mass_matrix, damping_matrix, stiffness_matrix):
    '''
    Analyze flutter conditions using system matrices.
    :param mass_matrix: Mass matrix of the structure.
    :param damping_matrix: Damping matrix of the structure.
    :param stiffness_matrix: Stiffness matrix of the structure.
    :return: Eigenvalues indicating stability.
    '''
    matrix_size = np.shape(mass_matrix)[0] * 2
    A = np.zeros((matrix_size, matrix_size))

    A[:matrix_size//2, matrix_size//2:] = np.eye(matrix_size//2)
    A[matrix_size//2:, :matrix_size//2] =
↪ -np.linalg.inv(mass_matrix).dot(stiffness_matrix)
    A[matrix_size//2:, matrix_size//2:] =
↪ -np.linalg.inv(mass_matrix).dot(damping_matrix)

    eigenvalues, _ = eig(A)
    return eigenvalues

# Example usage
density = 1.225  # Fluid density (kg/m^3)
free_stream_velocity = 15  # Free stream velocity (m/s)
diameter = 0.1  # Characteristic length (m)
area = 0.01  # Reference area (m^2)
lift_coefficient = 1.0  # Lift coefficient
time = np.linspace(0, 1, 100)  # Time array

# Calculate Strouhal frequency
strouhal_freq = strouhal_frequency(free_stream_velocity, diameter)
```

```
# Calculate lift force at different times
lift_forces = [lift_force_fluctuation(density, free_stream_velocity,
↪    area, lift_coefficient, strouhal_freq, t) for t in time]

print("Strouhal Frequency:", strouhal_freq)
print("Lift Forces:", lift_forces)

# Flutter analysis example matrix
mass_matrix = np.array([[1, 0], [0, 1]])
damping_matrix = np.array([[0.02, 0], [0, 0.02]])
stiffness_matrix = np.array([[10, 0], [0, 10]])

# Perform flutter analysis
eigenvalues = flutter_analysis(mass_matrix, damping_matrix,
↪    stiffness_matrix)

print("Flutter Eigenvalues:", eigenvalues)
```

This code defines several key functions essential for the analysis of flow-induced vibrations:

- `strouhal_frequency` function calculates the frequency of vortex shedding using the Strouhal number, providing insight into vortex-induced vibrations.

- `lift_force_fluctuation` computes the oscillating lift force on a structure, crucial for assessing the effect of fluid forces on structural dynamics.

- `flutter_analysis` determines the stability of a structure by evaluating its eigenvalues based on mass, damping, and stiffness matrices, aiding in flutter prediction.

The final block of code provides examples of these computations using hypothetical structural and fluid parameters.

Multiple Choice Questions

1. What drives the initiation of flow-induced vibrations (FIV) in structures?

    (a) External pressure fluctuations

    (b) Structural defects

    (c) Fluid-structure interaction

    (d) Material fatigue

2. Which of the following equations governs the dynamics of a fluid in the context of FIV?

    (a) Bernoulli's equation

    (b) Navier-Stokes equations

    (c) Continuity equation

    (d) Momentum integral equation

3. In the study of flow-induced vibrations, which term accounts for structural damping in the equation of motion for structures?

    (a) Mass matrix m

    (b) Stiffness matrix k

    (c) Damping matrix c

    (d) Fluid force $\mathbf{f}_{fluid}$

4. The Strouhal number St is crucial in analyzing which type of FIV mechanism?

    (a) Flutter

    (b) Vortex-induced vibrations

    (c) Buffeting

    (d) Laminar flow transition

5. What is the primary effect of flutter in engineering structures?

    (a) Energy dissipation

    (b) Random noise

    (c) Divergent oscillations

    (d) Resonance suppression

6. Buffeting is primarily associated with:

    (a) Steady aerodynamic forces

    (b) Unsteady aerodynamic forces

    (c) Constant lift coefficients

    (d) Uniform flow conditions

7. Which analytical approach is commonly employed to predict and mitigate FIV in complex systems?

    (a) Experimental testing
    (b) The finite element method and CFD
    (c) Simplified analytical models
    (d) Empirical correlation techniques

Answers:

1. **C: Fluid-structure interaction** Fluid-structure interaction is the driving force behind flow-induced vibrations, where fluid flow interacts dynamically with the structure.

2. **B: Navier-Stokes equations** The Navier-Stokes equations govern the fluid dynamics aspect of FIV, capturing momentum conservation in fluid flow.

3. **C: Damping matrix** c The damping matrix c in the equation of motion for structures accounts for energy dissipation due to damping in structural vibrations.

4. **B: Vortex-induced vibrations** The Strouhal number is used in the context of vortex-induced vibrations to relate vortex shedding frequency to flow and structural parameters.

5. **C: Divergent oscillations** Flutter can cause divergent oscillations due to the coupling of aerodynamic forces with structural modes, potentially leading to catastrophic failure.

6. **B: Unsteady aerodynamic forces** Buffeting results from unsteady aerodynamic forces caused by turbulent flow separations, affecting the dynamic response of structures.

7. **B: The finite element method and CFD** The finite element method (for structures) and computational fluid dynamics (CFD) are commonly used to predict and mitigate FIV by providing detailed insights into complex interactions between fluid flows and structural responses.

Chapter 25

Inviscid Irrotational Flow Theory

Theoretical Foundations of Inviscid Flow

The study of inviscid flows, wherein the viscosity of the fluid is neglected, simplifies the analysis of complex fluid dynamics problems by reducing the Navier-Stokes equations to Euler's equations. For inviscid flows, the foundational equation is the Euler equation, given by:

$$\rho \left(\frac{\partial \mathbf{u}}{\partial t} + \mathbf{u} \cdot \nabla \mathbf{u} \right) = -\nabla p \qquad (25.1)$$

where ρ denotes the fluid density, $\mathbf{u}$ is the velocity field, and p represents the pressure field. In inviscid flows, the absence of viscous forces allows a focus purely on pressure and inertial terms.

Irrotational Flow and Potential Function Theory

Irrotational flow, characterized by the absence of vorticity, $\nabla \times \mathbf{u} = \mathbf{0}$, permits the introduction of a velocity potential function, ϕ, where the velocity field is expressed as a gradient:

$$\mathbf{u} = \nabla \phi \qquad (25.2)$$

For irrotational flow, the velocity potential ϕ satisfies Laplace's equation:

$$\nabla^2 \phi = 0 \tag{25.3}$$

This implies that the potential function is harmonic, offering analytical solutions for simple boundary conditions and geometries.

Bernoulli's Equation in Potential Flow

In potential flow, Bernoulli's equation relates the velocity potential to pressure. For steady incompressible and irrotational flow, Bernoulli's equation is expressed as:

$$\frac{1}{2}\|\nabla\phi\|^2 + \frac{p}{\rho} + gz = \text{constant} \tag{25.4}$$

Here, $\|\nabla\phi\|$ denotes the magnitude of the velocity field, z the elevation head, and g the gravitational acceleration.

Stream Function and its Properties

For a two-dimensional irrotational flow, the stream function ψ is introduced such that:

$$u = \frac{\partial \psi}{\partial y}, \quad v = -\frac{\partial \psi}{\partial x} \tag{25.5}$$

The relation between ϕ and ψ for potential flow is governed by the Cauchy-Riemann equations, ensuring orthogonality and aiding in the construction of complex potentials for two-dimensional flow analysis.

Complex Potential and Conformal Mapping

Complex potential functions integrate the velocity potential and stream function to yield a comprehensive representation of flow in the complex plane. The complex potential $W(z)$ is expressed as:

$$W(z) = \phi + i\psi \tag{25.6}$$

with $z = x + iy$, the complex coordinate. Conformal mapping applies transformations in the complex plane to solve flow problems around complex geometries by conserving angles, crucial for aerodynamic shape analysis.

Flow Patterns and the Superposition Principle

Utilizing the superposition principle, one can construct diverse flow patterns by linear combination of elementary solutions, such as uniform flow, sources, sinks, and vortices. The linearity inherent in the Laplace equation permits such combinations, facilitating the modeling of complex flow configurations.

Applications in Engineering Analysis

Inviscid irrotational analysis finds application in several engineering domains, particularly in external flow problems where viscosity effects are negligible. Classic applications include lift and drag prediction on streamlined bodies, design of flow fields around airfoil sections, and conceptual development of potential flow past bluff bodies.

Python Code Snippet

Below is a Python code snippet that encompasses the fundamental computational elements of inviscid and irrotational flow theory, including the calculation of potential and stream functions, as well as the application of the superposition principle for flow pattern construction.

```
import numpy as np

def velocity_potential(x, y, strength, x_source, y_source):
    '''
    Calculate velocity potential for a source/sink of given
      strength.
    :param x: X-coordinate.
    :param y: Y-coordinate.
    :param strength: Strength of the source/sink.
    :param x_source: X-coordinate of the source/sink.
```

```python
    :param y_source: Y-coordinate of the source/sink.
    :return: Velocity potential value.
    '''
    return strength * np.log(np.sqrt((x - x_source)**2 + (y -
    ↪    y_source)**2))

def stream_function(x, y, strength, x_source, y_source):
    '''
    Calculate stream function for a source/sink of given strength.
    :param x: X-coordinate.
    :param y: Y-coordinate.
    :param strength: Strength of the source/sink.
    :param x_source: X-coordinate of the source/sink.
    :param y_source: Y-coordinate of the source/sink.
    :return: Stream function value.
    '''
    return strength * np.arctan2((y - y_source), (x - x_source))

def superposition_flow(x, y, source_sinks):
    '''
    Calculate the combined potential and stream function from
    ↪    multiple sources/sinks.
    :param x: X-coordinate grid.
    :param y: Y-coordinate grid.
    :param source_sinks: List of tuples defining strength and
    ↪    location of sources/sinks.
    :return: Combined velocity potential and stream function.
    '''
    phi_total = np.zeros_like(x)
    psi_total = np.zeros_like(x)

    for strength, x_s, y_s in source_sinks:
        phi_total += velocity_potential(x, y, strength, x_s, y_s)
        psi_total += stream_function(x, y, strength, x_s, y_s)

    return phi_total, psi_total

def flow_pattern_example():
    '''
    Example setup for superposition of flow patterns using source
    ↪    and sink.
    '''
    # Grid parameters
    x_range = np.linspace(-2, 2, 100)
    y_range = np.linspace(-2, 2, 100)
    x, y = np.meshgrid(x_range, y_range)

    # Define source/sink strengths and positions
    source_sinks = [(5.0, -1.0, 0.0), (-5.0, 1.0, 0.0)]

    # Calculate combined flow pattern
    phi, psi = superposition_flow(x, y, source_sinks)
```

```
    return x, y, phi, psi

# Run the flow pattern example
x, y, phi, psi = flow_pattern_example()
```

This code defines several essential functions for analyzing inviscid and irrotational flows:

- `velocity_potential` computes the velocity potential for point sources/sinks located in a fluid flow field.

- `stream_function` calculates the corresponding stream function, providing a visual representation of streamlines.

- `superposition_flow` enables the combination of multiple flow elements (sources/sinks) to form complex flow patterns through superposition.

- `flow_pattern_example` sets up a sample scenario combining velocity potential and stream functions to illustrate basic flow patterns from sources and sinks.

The final block demonstrates an example of creating a combined flow field using superposition principles to identify resultant potential and stream function across a grid.

Multiple Choice Questions

1. Which equation simplifies to the Euler equation under inviscid flow conditions?

    (a) Bernoulli's Equation

    (b) Navier-Stokes Equation

    (c) Continuity Equation

    (d) Reynolds-Averaged Navier-Stokes (RANS) Equation

2. For irrotational flow, which condition is true about the velocity field?

    (a) It can be expressed as a gradient of a scalar potential.

    (b) It contains non-zero vorticity.

    (c) It satisfies the Reynolds number.

(d) It can only be applied to compressible flows.

3. What does the stream function ψ represent in a two-dimensional flow?

   (a) The distribution of fluid density.

   (b) The magnitude of the velocity vector.

   (c) The velocity components orthogonal to the flow direction.

   (d) A scalar field whose contours represent streamlines.

4. Which one of these equations is satisfied by the velocity potential ϕ in irrotational flow?

   (a) Euler's Equation

   (b) Laplace's Equation

   (c) Navier-Stokes Equation

   (d) Bernoulli's Equation

5. In inviscid, irrotational flow, Bernoulli's equation relates pressure to:

   (a) Velocity only

   (b) Potential function and kinetic energy terms

   (c) Gravitational terms only

   (d) Vorticity and turbulence

6. The complex potential $W(z)$ is used in fluid dynamics to represent:

   (a) Combined effects of velocity and pressure gradients

   (b) The interaction between viscous and non-viscous forces

   (c) Unsteady flow patterns in incompressible fluids

   (d) Both the velocity potential and stream function in the complex plane

7. Conformal mapping is particularly useful in solving flow problems around:

   (a) Compressible media

   (b) Simplified geometries only

(c) Complex geometries by conserving angles

(d) One-dimensional flow fields

Answers:

1. **B: Navier-Stokes Equation** The Navier-Stokes equations, when the viscosity term is neglected, simplify to the Euler equations appropriate for inviscid flow analysis.

2. **A: It can be expressed as a gradient of a scalar potential.** In irrotational flow, the velocity field is derived from a scalar velocity potential, supporting calculation simplicity.

3. **D: A scalar field whose contours represent streamlines.** The stream function ψ is defined such that its level curves align with the fluid streamlines.

4. **B: Laplace's Equation** For irrotational flow, the velocity potential ϕ automatically satisfies Laplace's equation if the flow is incompressible.

5. **B: Potential function and kinetic energy terms** Bernoulli's equation in this context integrates both kinetic energy (through the potential function) and pressure, offering a comprehensive view of flow energy.

6. **D: Both the velocity potential and stream function in the complex plane** The complex potential $W(z) = \phi + i\psi$ offers a convenient way to pair and analyze these two aspects of flow simultaneously.

7. **C: Complex geometries by conserving angles** Conformal mapping maintains local angle properties, making it particularly powerful for solving flow problems around complex shapes.

Chapter 26

Mach Number and Flow Regimes

Introduction to the Mach Number

The Mach number, denoted as M, is a dimensionless quantity representing the ratio of flow velocity past a boundary to the local speed of sound:

$$M = \frac{U}{a}$$

where U is the flow velocity and a is the speed of sound in the medium. This pivotal parameter distinguishes between different flow regimes, dictating the behavior of compressible flows.

Speed of Sound and Thermodynamic Considerations

The speed of sound a in a medium depends on the thermodynamic properties of the fluid. For an ideal gas, it is defined as:

$$a = \sqrt{\gamma R T}$$

where γ is the specific heat ratio, R is the specific gas constant, and T is the absolute temperature. Understanding the interplay between these variables is essential for predicting changes in sonic conditions across varying pressures and temperatures.

Flow Regimes Classified by Mach Number

Several flow regimes exist based on the Mach number:
- Subsonic: $M < 1$ - Transonic: $M \approx 1$ - Supersonic: $M > 1$ - Hypersonic: $M \gg 1$

Each regime is characterized by distinct fluid behavior, particularly in terms of compressibility effects.

1 Subsonic Flow Dynamics

In subsonic flows, compressibility effects are typically negligible. The flow field is dominated by smooth and continuous changes in pressure, density, and velocity:

$$\nabla \cdot \mathbf{u} = 0$$

showing that incompressible assumptions can often apply to subsonic conditions without significant error.

2 Transonic Flow Characteristics

Transonic flow occurs as the Mach number approaches unity. This regime is significant due to the onset of shock waves and the coexistence of both subsonic and supersonic flow regions, making analysis complex. The potential equation:

$$\nabla^2 \phi + M^2 \nabla \phi \cdot \nabla \phi = 0$$

highlights the nonlinearity introduced by the compressible effects at transonic speeds.

3 Supersonic Flow and Shock Waves

In supersonic flows, the Mach number exceeds one, and shock waves become a dominant factor, leading to abrupt changes in flow properties. The governing equations transform to hyperbolic form, where the conservation laws are:

$$\frac{\partial \mathbf{Q}}{\partial t} + \nabla \cdot \mathbf{F}(\mathbf{Q}) = 0$$

accounting for nonlinear phenomena of shock waves and expansion fans.

4 Hypersonic Flow Observations

Hypersonic flows, characterized by very high Mach numbers, enhance the complexity due to chemical reactions, thermal nonequilibrium, and viscous interactions. The total enthalpy h_t becomes a critical parameter:

$$h_t = h + \frac{U^2}{2}$$

indicating the substantial kinetic energy impact on thermodynamic properties.

Applications and Numerical Methods

The application of Mach number in the design and optimization of high-speed vehicles, such as aircraft and rockets, requires specialized numerical methods. Techniques such as Computational Fluid Dynamics (CFD), using methods like finite volume and finite element, solve the governing equations across complex geometries. The iterative solvers and turbulence models must handle shock waves and boundary layer interactions, where:

$$\mathtt{residual} = \sum \left(\frac{\partial \mathbf{Q}}{\partial t} - \nabla \cdot \mathbf{F}(\mathbf{Q}) \right)$$

demonstrates the adjustment of the solution towards the convergence, offering insights into the fluid dynamics at different Mach numbers.

Python Code Snippet

Below is a Python code snippet that encompasses core computational elements in the analysis of Mach number and flow regimes, covering the calculation of Mach number, speed of sound, and classification of flow regimes with implications of specific flow features.

```
import numpy as np

def mach_number(U, a):
    '''
    Calculate the Mach number.
    :param U: Flow velocity in m/s.
    :param a: Speed of sound in m/s.
```

```python
    :return: Mach number.
    '''
    return U / a

def speed_of_sound(gamma, R, T):
    '''
    Calculate the speed of sound in an ideal gas.
    :param gamma: Specific heat ratio.
    :param R: Specific gas constant in J/(kg*K).
    :param T: Absolute temperature in Kelvin.
    :return: Speed of sound in m/s.
    '''
    return np.sqrt(gamma * R * T)

def flow_regime(M):
    '''
    Determine the flow regime based on the Mach number.
    :param M: Mach number.
    :return: String description of the flow regime.
    '''
    if M < 1:
        return "Subsonic"
    elif M == 1:
        return "Transonic"
    elif 1 < M < 5:
        return "Supersonic"
    else:
        return "Hypersonic"

def total_enthalpy(h, U):
    '''
    Calculate the total enthalpy in hypersonic flows.
    :param h: Static enthalpy in J/kg.
    :param U: Flow velocity in m/s.
    :return: Total enthalpy in J/kg.
    '''
    return h + 0.5 * U**2

# Example usage
U = 300  # Flow velocity in m/s
gamma = 1.4
R = 287.05  # Specific gas constant for air in J/(kg*K)
T = 288.15  # Absolute temperature in Kelvin

a = speed_of_sound(gamma, R, T)
M = mach_number(U, a)
regime = flow_regime(M)

h = 250000  # Example static enthalpy in J/kg
h_t = total_enthalpy(h, U)

print("Mach Number:", M)
print("Flow Regime:", regime)
```

```
print("Speed of Sound:", a)
print("Total Enthalpy:", h_t)
```

This code snippet defines several key functions necessary for analyzing Mach numbers and flow regimes:

- `mach_number` function computes the Mach number given the flow velocity and speed of sound.
- `speed_of_sound` calculates the speed of sound for an ideal gas based on its thermodynamic properties.
- `flow_regime` determines the flow regime according to the computed Mach number, distinguishing between subsonic, transonic, supersonic, and hypersonic flows.
- `total_enthalpy` computes the total enthalpy in hypersonic flows, taking into account significant kinetic energy contributions.

The final block of code demonstrates how to use these functions to perform calculations with typical atmospheric conditions and parameters.

Multiple Choice Questions

1. What is the Mach number primarily used to describe in fluid mechanics?

    (a) The pressure variation in a fluid system

    (b) The ratio of fluid velocity to gravitational force

    (c) The ratio of flow velocity to the local speed of sound

    (d) The density fluctuation in fluids

2. How is the speed of sound in a medium typically expressed for an ideal gas?

    (a) $a = \sqrt{\gamma P / \rho}$

    (b) $a = \gamma RT$

    (c) $a = \sqrt{\gamma RT}$

    (d) $a = \frac{P}{\rho}$

3. Which flow regime is characterized by having a Mach number much greater than one?

    (a) Subsonic
    (b) Transonic
    (c) Supersonic
    (d) Hypersonic

4. In a transonic flow regime, what significant feature commonly appears?

    (a) Laminar flow patterns
    (b) Smooth velocity distributions
    (c) Shock waves
    (d) Irrotational flows

5. Which of the following aspects is NOT significantly affected by compressibility in subsonic flows?

    (a) Pressure changes
    (b) Density fluctuations
    (c) Velocity distribution
    (d) Temperature changes

6. Which term in supersonic flow equations accounts for nonlinear phenomena like shock waves?

    (a) $\nabla \cdot \mathbf{u} = 0$
    (b) $\nabla^2 \phi$
    (c) $\frac{\partial \mathbf{Q}}{\partial t} + \nabla \cdot \mathbf{F}(\mathbf{Q}) = 0$
    (d) $\mathbf{Q} = \rho \mathbf{u}$

7. What is a key consideration when analyzing hypersonic flows?

    (a) Negligible kinetic energy
    (b) Chemical reactions and thermal nonequilibrium
    (c) Constant fluid density
    (d) Absence of shock waves

Answers:

1. **C: The ratio of flow velocity to the local speed of sound** The Mach number, denoted as M, is defined as the ratio of the flow velocity (U) to the local speed of sound (a) in the fluid.

2. **C: $a = \sqrt{\gamma RT}$** For an ideal gas, the speed of sound is given by $\sqrt{\gamma RT}$, where γ is the specific heat ratio, R is the specific gas constant, and T is the absolute temperature.

3. **D: Hypersonic** Hypersonic flows are characterized by very high Mach numbers ($M \gg 1$), typically greater than 5.

4. **C: Shock waves** In transonic regimes, shock waves are common due to the coexistence of subsonic and supersonic flow regions when the Mach number is approximately one.

5. **B: Density fluctuations** In subsonic flows, compressibility effects are often negligible, and thus density fluctuations are minimal.

6. **C: $\frac{\partial \mathbf{Q}}{\partial t} + \nabla \cdot \mathbf{F}(\mathbf{Q}) = 0$** In supersonic flows, the conservation equations are hyperbolic, dealing with nonlinear phenomena related to shock waves and expansion fans.

7. **B: Chemical reactions and thermal nonequilibrium** In hypersonic flows, complex effects such as chemical reactions and thermal nonequilibrium significantly affect fluid dynamics due to the high Mach numbers involved.

Chapter 27

Non-Newtonian Fluid Dynamics

Introduction to Non-Newtonian Fluids

Non-Newtonian fluid dynamics is a field concerned with fluids whose flow behavior deviates from the classical Newtonian fluid model. Unlike Newtonian fluids, characterized by a constant viscosity, non-Newtonian fluids exhibit a viscosity that can change under varying strain rates or shear stress. These fluids are essential in numerous engineering applications, from biomedical to chemical processing systems.

Theoretical Background and Governing Equations

The behavior of non-Newtonian fluids is often described by constitutive equations, relating stress and strain rate tensors. Key models describing these relationships are derived from empirical data and theoretical considerations. The general form of the constitutive relationship for a non-Newtonian fluid is given by:

$$\boldsymbol{\tau} = f(\dot{\gamma})$$

where $\boldsymbol{\tau}$ is the shear stress tensor, and $f(\dot{\gamma})$ represents the functional dependence on the shear rate tensor $\dot{\gamma}$.

1 Power-Law Model

The power-law model is a widely used empirical model for shear-thinning or shear-thickening fluids, where the viscosity η is expressed as:

$$\eta(\dot{\gamma}) = K\dot{\gamma}^{n-1}$$

where K is the flow consistency index, n is the flow behavior index, and $\dot{\gamma}$ is the shear rate. For $n < 1$, the fluid is shear-thinning, while for $n > 1$, it is shear-thickening.

2 Bingham Plastic Model

The Bingham plastic model describes a yield-stress fluid that behaves as a rigid body at low stresses but flows as a viscous fluid at high stresses. The relation is captured by:

$$\boldsymbol{\tau} = \tau_y + \eta_p \dot{\gamma}$$

where τ_y is the yield stress, and η_p is the plastic viscosity.

Rheological Measurement Techniques

Characterizing non-Newtonian fluids requires precise rheological measurement techniques to determine parameters such as viscosity, yield stress, and shear rates. Rheometers, designed to apply controlled stress or strain, are essential for this purpose.

1 Flow Curve Analysis

Flow curves plot shear stress versus shear rate, providing insights into fluid dynamics under varying conditions. The slope of these curves assists in calculating parameters n and K for various models, including the power-law.

Computational Fluid Dynamics Application

Non-Newtonian fluid mechanics are often simulated using computational fluid dynamics (CFD) to offer insights into fluid behavior under complex geometries and conditions. The Navier-Stokes equations, modified for non-Newtonian effects, form the backbone of such analyses:

$$\rho \left(\frac{\partial \mathbf{u}}{\partial t} + \mathbf{u} \cdot \nabla \mathbf{u} \right) = -\nabla p + \nabla \cdot \boldsymbol{\tau} + \mathbf{f}$$

where ρ is the fluid density, $\mathbf{u}$ is the velocity vector, p is the pressure, and $\mathbf{f}$ are body forces per unit volume.

1 Discretization Methods

Finite element and finite volume methods are typical discretization techniques employed in the numerical analysis of non-Newtonian fluids. These methods facilitate handling the complex boundary conditions and variable material properties inherent in non-Newtonian flows.

2 Turbulence Models for Non-Newtonian Fluids

Capturing the onset of turbulence in non-Newtonian fluid flows is challenging due to variable viscosity. Turbulence models such as the k-epsilon and k-omega are often adapted to account for the shear-dependent viscosity characteristics:

$$\nu_t = C_\mu \frac{k^2}{\epsilon} (\eta/\eta_0)^{n-1}$$

where ν_t is the turbulent viscosity, k is the turbulent kinetic energy, ϵ is the rate of dissipation, and η_0 is the reference viscosity.

Practical Engineering Applications

Non-Newtonian fluids have diverse engineering applications, ranging from polymer processing to the food industry. Understanding their complex flow behaviors under various conditions assists in the design of processes and equipment to optimize efficiency and performance.

1 Polymer Processing

In polymer processing, the understanding of non-Newtonian flow is crucial for predicting material behavior during extrusion and molding. The power-law and Carreau models often represent the viscosity profile of these polymeric materials.

2 Biological and Medical Applications

Non-Newtonian models help simulate blood flow, accounting for shear-thinning properties and pulsatile nature. These simulations assist in designing medical devices and understanding cardiovascular diagnostics.

Python Code Snippet

Below is a Python code snippet that encompasses the core computational elements for analyzing non-Newtonian fluid dynamics. This includes models for characterizing non-Newtonian behavior, simulation setup using CFD, and rheological measurements.

```python
import numpy as np

def power_law_viscosity(gamma_dot, K, n):
    '''
    Calculate viscosity for a power-law non-Newtonian fluid.
    :param gamma_dot: Shear rate.
    :param K: Flow consistency index.
    :param n: Flow behavior index.
    :return: Viscosity value.
    '''
    return K * (gamma_dot ** (n - 1))

def bingham_plastic_stress(gamma_dot, tau_y, eta_p):
    '''
    Calculate shear stress for a Bingham plastic fluid.
    :param gamma_dot: Shear rate.
    :param tau_y: Yield stress.
    :param eta_p: Plastic viscosity.
    :return: Shear stress value.
    '''
    return tau_y + eta_p * gamma_dot

def navier_stokes_non_newtonian(rho, u, p, tau, f):
    '''
    Simplified representation function for Navier-Stokes equation
    ↪ tailored for non-Newtonian fluids.
    :param rho: Fluid density.
    :param u: Velocity vector.
    :param p: Pressure.
    :param tau: Shear stress tensor.
    :param f: Body force vector.
    :return: Placeholder for momentum equation result.
    '''
    return rho * (np.gradient(u)[0] + u * np.gradient(u)[1]) == 
    ↪ -np.gradient(p) + np.gradient(tau) + f
```

```python
def calculate_flow_curve(gamma_dot_values, stress_func, *func_args):
    '''
    Generate a flow curve for a non-Newtonian fluid.
    :param gamma_dot_values: Array of shear rates.
    :param stress_func: Function to calculate shear stress.
    :param func_args: Additional arguments for stress function.
    :return: Array of shear stresses.
    '''
    return np.array([stress_func(gamma_dot, *func_args) for
        gamma_dot in gamma_dot_values])

def finite_element_simulation(mesh, material_properties,
    boundary_conditions):
    '''
    Placeholder for CFD simulation using finite element method.
    :param mesh: Mesh for the domain.
    :param material_properties: Non-Newtonian material properties.
    :param boundary_conditions: Applied boundary conditions.
    :return: Placeholder for simulation results.
    '''
    # Simulate the CFD solution (details are complex and typically
        involve specialized libraries like FEniCS)
    solution = "FEM simulation results"
    return solution

def adapted_turbulence_model_viscosity(nu_t, k, epsilon, eta, eta_0,
    n):
    '''
    Adjust turbulent viscosity for non-Newtonian effects in
        k-epsilon model.
    :param nu_t: Turbulent viscosity.
    :param k: Turbulent kinetic energy.
    :param epsilon: Dissipation rate.
    :param eta: Current viscosity.
    :param eta_0: Reference viscosity.
    :param n: Flow behavior index.
    :return: Adjusted turbulent viscosity.
    '''
    return nu_t * ((eta / eta_0) ** (n - 1))

# Example parameters for testing
gamma_dot_test = np.linspace(0, 10, 100)
K_test, n_test = 0.2, 0.8
tau_y_test, eta_p_test = 0.1, 0.3
rho_test, u_test, p_test, tau_test, f_test = 1.0, np.array([1.0,
    0.0]), 0.0, np.array([1.0, 0.0]), np.array([0.0, 0.0])

# Calculating viscosity using the power-law model
viscosity_values = power_law_viscosity(gamma_dot_test, K_test,
    n_test)

# Calculating shear stress using the Bingham plastic model
```

```
shear_stress_values = calculate_flow_curve(gamma_dot_test,
↪   bingham_plastic_stress, tau_y_test, eta_p_test)

# Example call to Navier-Stokes function (would be part of a larger
↪   CFD simulation framework)
momentum_equation = navier_stokes_non_newtonian(rho_test, u_test,
↪   p_test, tau_test, f_test)

# Placeholder CFD simulation (requires detailed setup in practice)
simulation_results = finite_element_simulation(None, None, None)

# Adjusted turbulent viscosity calculation
adjusted_nu_t = adapted_turbulence_model_viscosity(0.1, 1.0, 0.1,
↪   0.2, 0.2, 0.8)

print("Power-law Viscosity:", viscosity_values)
print("Bingham Plastic Shear Stress:", shear_stress_values)
print("Momentum Equation:", momentum_equation)
print("Simulation Results:", simulation_results)
print("Adjusted Turbulent Viscosity:", adjusted_nu_t)
```

This code provides essential functions and computations for handling non-Newtonian fluid dynamics:

- `power_law_viscosity` computes the viscosity of a power-law fluid.

- `bingham_plastic_stress` calculates the shear stress for a Bingham plastic fluid.

- `navier_stokes_non_newtonian` presents a simplified implementation of the Navier-Stokes equation for non-Newtonian fluids.

- `calculate_flow_curve` generates a flow curve (shear stress vs. shear rate) for analysis.

- `finite_element_simulation` is a placeholder for detailed CFD analysis using the finite element method.

- `adapted_turbulence_model_viscosity` adjusts turbulent viscosity values for non-Newtonian flow characteristics.

The script demonstrates calculations of viscosity and stress, a conceptual layout of a CFD setup, and adjustment methods for modeling turbulence in non-Newtonian fluids.

Multiple Choice Questions

1. Which of the following descriptions best fits non-Newtonian fluids?

    (a) Fluids with constant viscosity regardless of shear rate

    (b) Fluids characterized by temperature-dependent viscosity

    (c) Fluids with viscosity that varies with shear rate or shear stress

    (d) Ideal fluids with no viscosity

2. In the power-law model, what does the parameter n signify?

    (a) It represents the fluid density

    (b) It indicates the degree of shear-thinning or shear-thickening behavior

    (c) It defines the fluid's elasticity

    (d) It determines the compressibility of the fluid

3. For a Bingham plastic fluid, which of the following must be overcome for the fluid to start flowing?

    (a) Elasticity modulus

    (b) Plastic viscosity

    (c) Yield stress

    (d) Thermal expansion

4. What is the primary purpose of rheological measurement techniques in studying non-Newtonian fluids?

    (a) To measure the fluid's thermal conductivity

    (b) To determine the fluid's viscosity, yield stress, and shear rate

    (c) To calculate the fluid's pressure drop in a pipe

    (d) To measure the fluid's color and opacity

5. In computational fluid dynamics (CFD), how are non-Newtonian effects typically incorporated into the Navier-Stokes equations?

    (a) Through modifying the gravitational term

    (b) By adjusting the fluid density directly

    (c) By altering the shear stress term to account for variable viscosity

    (d) By ignoring viscous forces entirely

6. Which turbulence model is adapted for non-Newtonian fluid characteristics involving variable viscosity?

    (a) Laminar flow model

    (b) Eulerian flow model

    (c) Navier-Stokes model without modifications

    (d) k-epsilon or k-omega model with viscosity modification

7. Why is understanding non-Newtonian fluid dynamics crucial in polymer processing?

    (a) To predict color changes in polymers during processing

    (b) To account for the melting point variations in polymers

    (c) To accurately model and control viscosity variations during shaping processes

    (d) To analyze polymer thermal expansion in high-temperature conditions

Answers:

1. **C: Fluids with viscosity that varies with shear rate or shear stress** Non-Newtonian fluids are characterized by a viscosity that changes in response to the applied shear rate or shear stress, contrasting with Newtonian fluids where viscosity remains constant regardless of the shear rate.

2. **B: It indicates the degree of shear-thinning or shear-thickening behavior** In the power-law model, n is the flow behavior index signifying whether a fluid is shear-thinning ($n < 1$) or shear-thickening ($n > 1$).

3. **C: Yield stress** Bingham plastic fluids require an initial yield stress to be overcome before they behave as flowing viscous materials. Below this yield stress, they behave like rigid bodies.

4. **B: To determine the fluid's viscosity, yield stress, and shear rate** Rheological measurements are vital for characterizing properties such as viscosity, shear rate, and yield stress, which are crucial for understanding the flow behavior of non-Newtonian fluids.

5. **C: By altering the shear stress term to account for variable viscosity** In CFD simulations, non-Newtonian effects are incorporated by modifying the shear stress term in the Navier-Stokes equations to represent the variable viscosity characteristic of such fluids.

6. **D: k-epsilon or k-omega model with viscosity modification** These turbulence models are adapted to handle the variable viscosity in non-Newtonian fluids, allowing for more accurate predictions of flow behavior in turbulent conditions.

7. **C: To accurately model and control viscosity variations during shaping processes** In polymer processing, non-Newtonian fluid dynamics are essential for understanding how polymer melts flow and deform, crucial for process optimization and equipment design.

Chapter 28

Shock Wave and Expansion Fan Dynamics

Introduction to Supersonic Flows

Supersonic flows are characterized by flow velocities exceeding the speed of sound in the respective medium. These flows exhibit complex shock wave structures and expansion phenomena, governed by fundamental gas dynamics principles. As flow disturbances propagate at finite speeds, shock waves and expansion fans arise as mechanisms for altering pressure, density, and temperature discontinuities in supersonic conditions.

Shock Wave Fundamentals

Shock waves represent abrupt discontinuities within a flow field, marked by instantaneous changes in pressure, temperature, and density. The Rankine-Hugoniot relations encapsulate these changes across the shock:

$$\frac{\rho_2}{\rho_1} = \frac{(+1)M_1^2}{(-1)M_1^2 + 2}$$

$$\frac{p_2}{p_1} = 1 + \frac{2}{+1}\left(M_1^2 - 1\right)$$

$$\frac{T_2}{T_1} = \frac{\left[\left(1 + \frac{2(M_1^2-1)}{+1}\right)\right](+1)M_1^2}{(-1)M_1^2 + 2}$$

where ρ denotes the density, p the pressure, T the temperature, M the Mach number, and γ the specific heat ratio. Indices 1 and 2 reference conditions upstream and downstream of the shock, respectively.

1 Normal and Oblique Shock Waves

In supersonic aerodynamics, shock waves are categorized as normal or oblique based on their inclination to the flow direction. For normal shocks, the entire downstream region experiences subsonic flow, ruled strictly by the aforementioned Rankine-Hugoniot relations. Oblique shocks, however, present an additional deflection angle θ satisfying:

$$\tan(\theta) = 2\cot(\beta)\left(\frac{M_1^2\sin^2(\beta) - 1}{M_1^2(\gamma + \cos(2\beta)) + 2}\right)$$

where β is the shock angle between the initial flow direction and the shock wave front.

Expansion Fan Dynamics

When a supersonic flow experiences an expansion around a corner, the Prandtl-Meyer expansion fan emerges as a series of infinitesimal Mach waves. The Prandtl-Meyer function $\nu(M)$ for an isentropic expansion is expressed as:

$$\nu(M) = \sqrt{\frac{+1}{-1}}\arctan\left(\sqrt{\frac{-1}{+1}(M^2-1)}\right) - \arctan\left(\sqrt{M^2-1}\right)$$

The turn angle θ in expansion fans can be related to Mach values before and after the expansion by:

$$\Delta\theta = \nu(M_2) - \nu(M_1)$$

where $\Delta\theta$ is the angular change, and M_1, M_2 denote pre- and post-expansion Mach numbers, respectively.

1 Mach Waves and Characteristics

Infinitesimally small perturbations in supersonic flows are dictated by Mach waves, aligning themselves at the Mach angle μ, given as:

$$\mu = \arcsin\left(\frac{1}{M}\right)$$

These angles define the minimum disturbance propagation pathways within the flow field and are fundamental to the understanding of linearization capabilities in supersonic flow analysis.

Mathematical Formulations in Shock-Expansion Theory

The interaction of shock waves and expansion fans is paramount in the design of supersonic nozzles and inlets. The shock-expansion theory employs conservation laws, merged with characteristic line methods, forming a rigorous framework:

$$\text{Continuity:} \quad \frac{\partial \rho}{\partial t} + \nabla \cdot (\rho \mathbf{u}) = 0$$

$$\text{Momentum:} \quad \frac{\partial (\rho \mathbf{u})}{\partial t} + \nabla \cdot (\rho \mathbf{u}\mathbf{u} + p\mathbf{I}) = 0$$

$$\text{Energy:} \quad \frac{\partial E}{\partial t} + \nabla \cdot ((E + p)\mathbf{u}) = 0$$

where ρ denotes density, $\mathbf{u}$ the velocity vector, p the pressure, $\mathbf{I}$ the identity matrix, and E the total energy per unit volume. These governing equations inform numerical simulation methodologies, critical in computational fluid dynamics studies.

1 CFD Implementation

In computational models, the discretization of governing equations often utilizes finite volume methods (FVM), prominently featuring shock-capturing schemes like Monotone Upstream-centered Schemes for Conservation Laws (MUSCL) or higher-order methods such as Weighted Essentially Non-Oscillatory (WENO) schemes. These approaches accurately resolve steep gradient regions with minimized numerical dissipation, maintaining stability across shock discontinuities.

Applications in Supersonic Engineering

Shock wave and expansion fan interactions are critical in designing supersonic aircraft, such as Concorde or the SR-71 Blackbird, optimizing aerodynamic efficiency. Additionally, the shock-expansion method determines nozzle contours in rocket engines, maximizing thrust by aligning exit conditions to free-stream Mach numbers.

Python Code Snippet

Below is a Python code snippet that encompasses the core computational elements related to shock wave and expansion fan dynamics, including the calculation of shock properties, expansion fan characteristics, and the implementation of computational fluid dynamics schemes.

```python
import numpy as np

def calculate_shock_properties(M1, gamma):
    '''
    Calculate the properties of flow across a normal shock wave
        using Rankine-Hugoniot relations.
    :param M1: Mach number before the shock.
    :param gamma: Specific heat ratio.
    :return: Density ratio, Pressure ratio, Temperature ratio.
    '''
    density_ratio = ((gamma + 1) * M1**2) / ((gamma - 1) * M1**2 +
        2)
    pressure_ratio = 1 + (2 * gamma / (gamma + 1)) * (M1**2 - 1)
    temperature_ratio = (((1 + ((2 * gamma * (M1**2 - 1)) / (gamma +
        1))) * (gamma + 1) * M1**2) /
                        ((gamma - 1) * M1**2 + 2))

    return density_ratio, pressure_ratio, temperature_ratio

def prandtl_meyer_function(M, gamma):
    '''
    Compute the Prandtl-Meyer function for expansion fans.
    :param M: Mach number.
    :param gamma: Specific heat ratio.
    :return: Prandtl-Meyer function value.
    '''
    term1 = np.sqrt((gamma + 1) / (gamma - 1))
    term2 = np.arctan(np.sqrt((gamma - 1) / (gamma + 1) * (M**2 -
        1)))
    term3 = np.arctan(np.sqrt(M**2 - 1))
```

```python
        return term1 * term2 - term3

def mach_wave_angle(M):
    '''
    Calculate the Mach wave angle.
    :param M: Mach number.
    :return: Mach angle in radians.
    '''
    return np.arcsin(1 / M)

def shock_capture_scheme(gamma, courant_number, grid_points):
    '''
    Simulated implementation of a shock-capturing method using
    ↪  finite volume approach.
    :param gamma: Specific heat ratio.
    :param courant_number: Courant number for stability.
    :param grid_points: Number of grid points in the simulation.
    :return: Simulated result (placeholder).
    '''
    # Placeholder grid initialization
    rho = np.ones(grid_points)
    u = np.zeros(grid_points)
    p = np.ones(grid_points)
    E = np.ones(grid_points)

    # Time-stepping loop (dummy implementation)
    for step in range(100):  # Fixed number of steps
        # Calculate fluxes (simplified)
        flux_rho = rho * u
        flux_u = rho * u**2 + p
        flux_E = (E + p) * u

        # Update (dummy implementation for demonstration)
        rho = rho - courant_number * (np.roll(flux_rho, -1) -
        ↪  flux_rho)
        u = u - courant_number * (np.roll(flux_u, -1) - flux_u)
        E = E - courant_number * (np.roll(flux_E, -1) - flux_E)

    return rho, u, p, E

# Example usage of these functions
Mach_number = 2.5
gamma_value = 1.4

rho_ratio, p_ratio, T_ratio =
↪  calculate_shock_properties(Mach_number, gamma_value)
prandtl_meyer_angle = prandtl_meyer_function(Mach_number,
↪  gamma_value)
mach_angle = mach_wave_angle(Mach_number)
simulation_result = shock_capture_scheme(gamma_value, 0.5, 100)

print("Density Ratio:", rho_ratio)
print("Pressure Ratio:", p_ratio)
```

```
print("Temperature Ratio:", T_ratio)
print("Prandtl-Meyer Angle:", prandtl_meyer_angle)
print("Mach Angle (radians):", mach_angle)
```

This code defines several key functions necessary for the modeling and understanding of shock wave and expansion fan dynamics:

- `calculate_shock_properties` computes the changes in flow properties due to a normal shock wave using Rankine-Hugoniot relations.

- `prandtl_meyer_function` evaluates the Prandtl-Meyer function, crucial for determining flow properties in expansion fans.

- `mach_wave_angle` calculates the Mach angle, defining the direction along which infinitesimal pressure disturbances propagate.

- `shock_capture_scheme` simulates a basic finite volume shock-capturing method for demonstrating computational fluid dynamics concepts.

The final block of code showcases the application of these functions to calculate important properties and setup a dummy simulation for educational purposes.

Multiple Choice Questions

1. Which of the following best describes a shock wave in a supersonic flow?

    (a) A gradual change in density and temperature.

    (b) An abrupt discontinuity within the flow field.

    (c) A flow regime with continuous expansion.

    (d) A region where velocity is minimized.

2. What does the Rankine-Hugoniot relation primarily describe?

    (a) The continuity in subsonic flow.

    (b) The energy conservation across a flow.

    (c) Discontinuities in pressure, density, and temperature across a shock wave.

(d) A smooth transition between subsonic and supersonic flow.

3. In the context of oblique shock waves, what does θ represent?

    (a) Deflection angle of the flow direction.
    (b) The shock angle relative to the oncoming flow.
    (c) The angular velocity of the shock.
    (d) The angle between Mach waves.

4. The Prandtl-Meyer function is associated with which of the following phenomena?

    (a) Shock waves.
    (b) Subsonic transition.
    (c) Compressibility effects.
    (d) Expansion fans.

5. Which equation is used to define the relationship between Mach waves in an expansion fan?

    (a) The Euler equation.
    (b) $\mu = \arcsin\left(\frac{1}{M}\right)$
    (c) Bernoulli's equation.
    (d) The Mach line equation.

6. In computational fluid dynamics for supersonic flows, which method is commonly used to capture shocks accurately?

    (a) Finite Element Method (FEM).
    (b) Crank-Nicolson scheme.
    (c) Monotone Upstream-centered Schemes for Conservation Laws (MUSCL).
    (d) Runge-Kutta Methods.

7. In supersonic engineering applications, why are shock waves and expansion fans particularly significant?

    (a) They determine the color profile of the flow.
    (b) They affect the thermal conductivity of the medium.

(c) They influence aerodynamic efficiency and thrust in supersonic aircraft and rockets.

(d) They are critical in heating the fluid to desired temperatures.

Answers:

1. **B: An abrupt discontinuity within the flow field** Shock waves are characterized by rapid changes in flow properties like pressure and temperature, creating abrupt flow discontinuities.

2. **C: Discontinuities in pressure, density, and temperature across a shock wave** The Rankine-Hugoniot relations describe changes in flow properties like pressure, density, and temperature that occur across shock waves.

3. **A: Deflection angle of the flow direction** In oblique shock waves, θ represents the angle by which the flow direction is deflected when passing through the shock wave.

4. **D: Expansion fans** The Prandtl-Meyer function describes isentropic expansion processes, specifically associated with the expansion fan phenomena in supersonic flows.

5. **B:** $\mu = \arcsin\left(\frac{1}{M}\right)$ This equation describes the Mach angle, which defines the alignment of Mach waves in a supersonic flow, crucial for expansion fan analysis.

6. **C: Monotone Upstream-centered Schemes for Conservation Laws (MUSCL)** This shock-capturing scheme is widely used in CFD to accurately model and simulate shock waves without introducing excessive numerical diffusion.

7. **C: They influence aerodynamic efficiency and thrust in supersonic aircraft and rockets** Shock waves and expansion fans are critical in determining the aerodynamic characteristics and performance in supersonic engineering applications, such as in aircraft and rockets.

Chapter 29

Fluid Flow in Porous Media

Governing Equations for Porous Media Flow

The study of fluid flows through porous media is foundational for understanding numerous applications in civil and geotechnical engineering. The governing equations in this context originate from the principles of mass conservation and Darcy's law, which describes the flow in porous environments.

The continuity equation for incompressible fluid flow is given by:

$$\nabla \cdot \mathbf{v} = 0$$

where $\mathbf{v}$ is the macroscopic velocity vector of the fluid.

Darcy's law forms the basis for quantifying fluid flow through porous media, expressed as:

$$\mathbf{q} = -\frac{\mathbf{K}}{\mu}\nabla P$$

where $\mathbf{q}$ is the Darcy flux, $\mathbf{K}$ is the permeability tensor reflecting the anisotropic nature of the porous medium, μ is the dynamic viscosity of the fluid, and P is the pressure potential driving the flow.

Incorporating Darcy's law into the mass conservation principle yields the fundamental equation of flow through porous media:

$$\nabla \cdot \left(-\frac{\mathbf{K}}{\mu}\nabla P\right) = 0$$

Permeability and Porosity Analysis

Permeability, $\mathbf{K}$, is a critical property of porous media, highlighting the ease with which fluids can traverse the medium's structure. It depends on the pore structure and connectivity, which are influenced by the media's geotechnical composition.

The porosity ϕ of a material is defined by:

$$\phi = \frac{V_v}{V_t}$$

where V_v represents the void volume and V_t the total volume. Porosity impacts the fluid storage capacity and affects flow dynamics within the medium.

Empirical correlations and permeability models, such as those proposed by Kozeny-Carman, relate permeability to porosity and specific surface area:

$$\mathbf{K} = \frac{\phi^3}{S^2(1-\phi)^2}$$

where S represents the specific surface area per unit volume of the solid matrix.

Application in Civil and Geotechnical Engineering

In geotechnical engineering, flow through porous media is essential for assessing soil stability, seepage through dams, and groundwater movement. The seepage velocity $\mathbf{v_s}$ is derived from Darcy's velocity by:

$$\mathbf{v_s} = \frac{\mathbf{q}}{\phi}$$

In civil engineering, understanding the interaction between fluids and subterranean formations aids in the design of foundations, tunnels, and retaining structures.

For the analysis of aquifer systems, the governing flow equations adapt to transient conditions with the storage coefficient S:

$$S\frac{\partial h}{\partial t} = \nabla \cdot (\mathbf{K}\nabla h) + W$$

where h is the hydraulic head and W represents sources or sinks within the aquifer.

Numerical Modeling in Porous Media

Solving flow equations in complex porous media geometries often necessitates numerical methods, prominently finite element (FE) and finite difference (FD) approaches.

The finite element method discretizes the spatial domain using elements, permitting detailed resolution of heterogeneous permeabilities. The weak form of the governing equations is expressed as:

$$\int_\Omega \nabla w \cdot \mathbf{K}\nabla P \, d\Omega = \int_\Omega w \, q \, d\Omega$$

where w is a weighting function.

Finite difference methods apply discretization to both time and space, providing explicit or implicit schemes for temporal evolution. For example, applying a backward difference scheme leads to:

$$\frac{h^{n+1} - h^n}{\Delta t} = \frac{1}{S}\nabla \cdot \left(\mathbf{K}\nabla h^{n+1}\right)$$

Advanced Topics in Reactive and Multi-phase Flows

Beyond single-phase flows, porous media flow encompasses reactive and multiphase systems. Reactive transport models incorporate chemical reactions between species, using:

$$\nabla \cdot (\mathbf{q}C) = \nabla \cdot (D\nabla C) + R$$

where C denotes concentration, D the dispersion tensor, and R reaction rate terms.

Further complexities arise in multiphase flow, often represented by phase-dependent Darcy's laws and capillary pressure-saturation relationships, capturing the interaction between liquid, gas, and solid phases.

Python Code Snippet

Below is a Python code snippet that encompasses the core computational elements for fluid flow through porous media including Darcy's law application, permeability and porosity analysis, numerical modeling, and consideration of multiphase and reactive flows.

```python
import numpy as np
from scipy.sparse import diags
from scipy.sparse.linalg import spsolve

def darcy_flux(K, mu, P_grad):
    '''
    Calculate the Darcy flux for flow through porous media.
    :param K: Permeability tensor.
    :param mu: Dynamic viscosity.
    :param P_grad: Pressure gradient.
    :return: Darcy flux.
    '''
    return -K / mu * P_grad

def porosity(volatile_volume, total_volume):
    '''
    Calculate the porosity of a material.
    :param volatile_volume: Volume of void spaces.
    :param total_volume: Total volume of the material.
    :return: Porosity.
    '''
    return volatile_volume / total_volume

def kozeny_carman(K, phi, S):
    '''
    Estimate permeability using Kozeny-Carman relation.
    :param K: Initial permeability value.
    :param phi: Porosity.
    :param S: Specific surface area.
    :return: Modified permeability.
    '''
    return phi**3 / (S**2 * (1 - phi)**2)

def seepage_velocity(flux, porosity):
    '''
    Calculate seepage velocity from Darcy's velocity.
    :param flux: Darcy flux.
    :param porosity: Porosity of the medium.
```

```python
    :return: Seepage velocity.
    '''
    return flux / porosity

def transient_aquifer_flow(K, S, h_initial, dt, dx, source_sink):
    '''
    Compute transient groundwater flow in an aquifer using finite
    ↪ difference method.
    :param K: Hydraulic conductivity.
    :param S: Storativity.
    :param h_initial: Initial head distribution.
    :param dt: Time step.
    :param dx: Space increment.
    :param source_sink: Source or sink term.
    :return: Head distribution after one time step.
    '''
    n = len(h_initial)
    alpha = K * dt / (S * dx**2)
    main_diag = (1 + 2 * alpha) * np.ones(n)
    off_diag = -alpha * np.ones(n-1)
    A = diags([off_diag, main_diag, off_diag], [-1, 0, 1]).tocsc()

    b = h_initial + (dt / S) * source_sink

    return spsolve(A, b)

def simulate_reactive_transport(q, C, D, R, dt, dx):
    '''
    Simulate reactive transport in porous media.
    :param q: Darcy flux.
    :param C: Initial concentration distribution.
    :param D: Dispersion coefficient.
    :param R: Reaction rate term.
    :param dt: Time step.
    :param dx: Space increment.
    :return: Concentration distribution after one time step.
    '''
    n = len(C)
    alpha = D * dt / dx**2
    advect = q * dt / dx
    main_diag = (1 + 2 * alpha + advect) * np.ones(n)
    off_diag = -alpha * np.ones(n-1)
    A = diags([off_diag, main_diag, off_diag], [-1, 0, 1]).tocsc()

    b = C + R * dt

    return spsolve(A, b)

# Example parameters
K_value = 1e-12
mu_value = 0.001
P_grad = np.array([100, 80])
vol_volume = 0.2
```

```
tot_volume = 1.0
S_value = 0.01
initial_head = np.array([10, 10, 10, 10, 10])
source_sink_term = np.zeros_like(initial_head)
dt_example = 0.01
dx_example = 0.1

# Example calculation
flux = darcy_flux(K_value, mu_value, P_grad)
phi = porosity(vol_volume, tot_volume)
modified_K = kozeny_carman(K_value, phi, S_value)
seepage_vel = seepage_velocity(flux, phi)
head_dist = transient_aquifer_flow(K_value, S_value, initial_head,
    dt_example, dx_example, source_sink_term)

print("Darcy Flux:", flux)
print("Porosity:", phi)
print("Modified Permeability:", modified_K)
print("Seepage Velocity:", seepage_vel)
print("Head Distribution:", head_dist)
```

This code defines several key functions necessary for understanding fluid flows through porous media:

- `darcy_flux` calculates the fluid flux in porous media using Darcy's law.

- `porosity` determines material porosity from volumetric parameters.

- `kozeny_carman` estimates permeability based on porosity and surface area using the Kozeny-Carman relation.

- `seepage_velocity` computes seepage velocity from Darcy's flux and porosity.

- `transient_aquifer_flow` models transient flow conditions in aquifers through finite difference methods.

- `simulate_reactive_transport` simulates concentration changes over time within a reactive transport system.

The final block of code illustrates example parameter values and performs various calculations to demonstrate the application of these functions.

Multiple Choice Questions

1. Which principle serves as the foundation for the governing equations of flow through porous media?

    (a) Bernoulli's principle

    (b) Navier-Stokes equations

    (c) Darcy's law and mass conservation

    (d) Euler's equations

2. In the context of porous media, what does the parameter **K** represent?

    (a) Dynamic viscosity of the fluid

    (b) Pressure potential

    (c) Permeability of the porous medium

    (d) Darcy flux

3. The expression $\phi = \frac{V_v}{V_t}$ defines which property of a porous medium?

    (a) Permeability

    (b) Porosity

    (c) Darcy flux

    (d) Specific surface area

4. What is the role of the specific surface area S in the Kozeny-Carman equation for permeability?

    (a) It represents the total pore volume.

    (b) It affects the macroscopic velocity.

    (c) It defines the pressure gradient.

    (d) It is used to relate permeability to porosity.

5. How is seepage velocity $\mathbf{v_s}$ in porous media calculated?

    (a) $\mathbf{v_s} = \frac{\phi}{q}$

    (b) $\mathbf{v_s} = \mathbf{q} \times \phi$

    (c) $\mathbf{v_s} = \frac{q}{\phi}$

    (d) $\mathbf{v_s} = \mathbf{q} + \phi$

6. In numerical modeling of porous media, what is a primary advantage of the finite element method?

    (a) It provides an explicit temporal scheme.
    (b) It requires less computational resource than finite difference methods.
    (c) It offers detailed resolution of heterogeneous permeabilities.
    (d) It uses backward difference schemes for time discretization.

7. Which factors contribute to complexities in modeling multiphase flow in porous media?

    (a) Chemical reactions and capillary pressure-saturation relationships
    (b) Simplified one-phase flow equations
    (c) Constant permeability tensor
    (d) Uniform porosity throughout the medium

Answers:

1. **C: Darcy's law and mass conservation** Darcy's law, combined with the principle of mass conservation, serves as the foundation for deriving the equations that govern fluid flow through porous media.

2. **C: Permeability of the porous medium K** denotes the permeability tensor, which quantifies the ability of the porous medium to allow fluid flow.

3. **B: Porosity** This formula defines porosity, which measures the fraction of void space in a material, essential for determining fluid storage capacity.

4. **D: It is used to relate permeability to porosity** In the Kozeny-Carman equation, the specific surface area S helps establish a relationship between permeability and porosity, affecting the ease of fluid flow.

5. **C: $\mathbf{v_s} = \frac{\mathbf{q}}{\phi}$** Seepage velocity $\mathbf{v_s}$ is determined by dividing the Darcy velocity (or flux) $\mathbf{q}$ by the porosity ϕ, accounting for flow through the porous medium.

6. **C: It offers detailed resolution of heterogeneous permeabilities** The finite element method excels in handling complex geometries and material properties, providing high-resolution analysis of flows through varying permeabilities.

7. **A: Chemical reactions and capillary pressure-saturation relationships** These factors introduce complexities in multiphase flow models, requiring advanced formulations to capture interactions between phases and reactions.

Chapter 30

Heat Exchanger Performance Equations

Introduction

Heat exchangers are critical components in various mechanical engineering applications, facilitating the transfer of thermal energy between two or more fluid streams. The performance of heat exchangers is governed by a set of mathematical equations that describe the fluid dynamics and temperature profiles within the system. Advanced models incorporate not only simple energy balances but also account for complex interactions involving heat transfer coefficients and fluid properties.

Governing Equations

The fundamental energy balance equation for a differential control volume in a heat exchanger can be expressed as:

$$dQ = \dot{m} c_p dT$$

where dQ is the differential heat transfer rate, $\dot{m}$ represents the mass flow rate, and c_p is the specific heat capacity at constant pressure. This relation is crucial in calculating the temperature variation along the length of the heat exchanger.

1 The Log Mean Temperature Difference (LMTD)

The LMTD method is a conventional approach used to approximate the average temperature difference between the hot and cold fluids across the heat exchanger. The LMTD is defined as:

$$\Delta T_{\text{lm}} = \frac{\Delta T_1 - \Delta T_2}{\ln\left(\frac{\Delta T_1}{\Delta T_2}\right)}$$

where ΔT_1 and ΔT_2 are the temperature differences at each end of the heat exchanger. This equation assumes a constant heat transfer coefficient and is particularly useful for preliminary design and analysis.

2 Effectiveness-NTU Method

The effectiveness-NTU (Number of Transfer Units) method offers a more detailed model by relating the heat exchanger effectiveness ε to the NTU, which is defined as:

$$\text{NTU} = \frac{UA}{C_{\min}}$$

Here, U denotes the overall heat transfer coefficient, A is the heat transfer area, and $C_{\min}$ is the minimum heat capacity rate of the two fluid streams. The effectiveness ε determines the performance and is expressed as:

$$\varepsilon = \frac{Q_{\text{actual}}}{Q_{\max}}$$

Heat Transfer Coefficients

The heat transfer coefficients for both the shell and tube sides are fundamental in determining the overall transfer efficiency of a heat exchanger. These coefficients are evaluated using empirical correlations, typically functions of Reynolds (Re), Prandtl (Pr), and Nusselt (Nu) numbers. The Dittus-Boelter equation provides a standard expression for turbulent flow inside tubes:

$$\text{Nu} = 0.023 \, \text{Re}^{0.8} \, \text{Pr}^n$$

where $n = 0.4$ and $n = 0.3$ for heating and cooling, respectively. The resulting Nusselt number allows the calculation of the convective heat transfer coefficient, h, given by:

$$h = \frac{\text{Nu} \cdot k}{d}$$

with k being the thermal conductivity of the fluid and d the characteristic dimension of the flow.

Modeling Temperature Profiles

In heat exchangers, accurately predicting temperature profiles involves analyzing the variations along the flow paths of the fluids. Advanced models incorporate axial conduction effects and variable properties. The temperature difference between the fluids can be approximated using integration techniques over the exchanger length, yielding:

$$T_{\text{out}} = T_{\text{in}} \pm \int_0^L \frac{q}{\dot{m}c_p}\,dx$$

where q is the local heat flux, and L is the length of the exchanger.

1 Counterflow and Parallel Flow Configurations

The spatial configuration of heat exchangers alters their performance characteristics. In counterflow arrangements, fluids flow in opposite directions, maximizing the temperature difference across the exchanger. For a counterflow heat exchanger, the effectiveness is maximized, expressed by the relation:

$$\Delta T_{\text{counter}} > \Delta T_{\text{parallel}}$$

This principle underscores the significance of flow arrangement in optimizing heat transfer.

Advanced Approaches and Computational Methods

The increasing demand for precision in thermal systems has prompted the integration of computational fluid dynamics (CFD) in heat exchanger analysis. These advanced simulations provide detailed insights into local temperature and velocity fields, accommodating complex geometries and varying material properties.

1 Utilization of CFD in Heat Exchanger Design

CFD tools enable the evaluation of intricate flow and temperature patterns, offering a detailed prediction of heat transfer performance. The governing equations for mass, momentum, and energy are solved numerically, often requiring iterative methods and turbulence models, such as k-epsilon or Large Eddy Simulation (LES), to capture the nuanced behavior within the exchanger.

2 Optimization Techniques

The optimization of heat exchanger design parameters can be facilitated via computational algorithms, enhancing efficiency and reducing thermal resistance. Techniques like genetic algorithms or gradient-based methods iterate design variables to maximize performance indices, such as minimization of entropy generation or maximization of heat recovery.

Python Code Snippet

Below is a Python code snippet that encompasses the core computational elements of heat exchanger performance equations, including the calculation of LMTD, effectiveness-NTU method, and heat transfer coefficients.

```
import numpy as np

def lmtd(delta_T1, delta_T2):
    '''
    Calculate the Log Mean Temperature Difference (LMTD).
    :param delta_T1: Temperature difference at one end.
    :param delta_T2: Temperature difference at the other end.
    :return: LMTD value.
    '''
    if delta_T1 == delta_T2:
        return delta_T1  # To avoid division by zero when the
            differences are equal
    else:
        return (delta_T1 - delta_T2) / np.log(delta_T1 / delta_T2)
```

```python
def ntu_method(U, A, C_min):
    '''
    Compute the Number of Transfer Units (NTU).
    :param U: Overall heat transfer coefficient.
    :param A: Heat transfer area.
    :param C_min: Minimum heat capacity rate.
    :return: NTU value.
    '''
    return (U * A) / C_min

def effectiveness_ntu(epsilon, ntu, C_min, C_max):
    '''
    Compute the heat exchanger effectiveness using the NTU method.
    :param epsilon: Effectiveness.
    :param ntu: Number of Transfer Units.
    :param C_min: Minimum heat capacity rate.
    :param C_max: Maximum heat capacity rate.
    :return: Actual heat transfer rate.
    '''
    if C_min == C_max:
        # Simplified case when both capacity rates are equal
        return ntu / (1 + ntu)
    else:
        retur_return = 1 - np.exp((-1) * ntu * ((1 - epsilon) /
        ↪ epsilon))
        return np.maximum(0, np.minimum(1, retur_return))

def heat_transfer_coefficient(Re, Pr, k, d, heating=True):
    '''
    Calculate convective heat transfer coefficient.
    :param Re: Reynolds number.
    :param Pr: Prandtl number.
    :param k: Thermal conductivity.
    :param d: Diameter of the pipe.
    :param heating: Boolean, True if heating, False if cooling.
    :return: Convective heat transfer coefficient.
    '''
    n = 0.4 if heating else 0.3  # different constant for heating
    ↪ and cooling
    Nu = 0.023 * (Re ** 0.8) * (Pr ** n)
    return (Nu * k) / d

def temperature_profile(T_in, q, m_dot, c_p, L, dx):
    '''
    Compute the outlet temperature along a heat exchanger.
    :param T_in: Inlet temperature.
    :param q: Local heat flux.
    :param m_dot: Mass flow rate.
    :param c_p: Specific heat capacity.
    :param L: Length of the heat exchanger.
    :param dx: Differential element in the length.
    :return: Outlet temperature.
    '''
```

```python
    T_out = T_in
    for _ in np.arange(0, L, dx):
        T_out += (q / (m_dot * c_p)) * dx
    return T_out

# Example usage and outputs
delta_T1_example = 50
delta_T2_example = 30
U_example = 500
A_example = 50
C_min_example = 2000
Re_example = 20000
Pr_example = 0.7
k_example = 0.6
d_example = 0.05
T_in_example = 300
q_example = 10000
m_dot_example = 5
c_p_example = 4.18
L_example = 10
dx_example = 0.1

lmtd_value = lmtd(delta_T1_example, delta_T2_example)
ntu_value = ntu_method(U_example, A_example, C_min_example)
effectiveness = effectiveness_ntu(0.75, ntu_value, C_min_example,
↪  C_min_example * 1.5)
h_coefficient = heat_transfer_coefficient(Re_example, Pr_example,
↪  k_example, d_example)
T_out_example = temperature_profile(T_in_example, q_example,
↪  m_dot_example, c_p_example, L_example, dx_example)

print("LMTD:", lmtd_value)
print("NTU:", ntu_value)
print("Effectiveness:", effectiveness)
print("Heat Transfer Coefficient:", h_coefficient)
print("Outlet Temperature:", T_out_example)
```

This Python code defines several key functions necessary for evaluating the performance of heat exchangers:

- lmtd computes the Log Mean Temperature Difference, which is crucial for approximating average temperature differences in heat exchangers.

- ntu_method calculates the Number of Transfer Units, helping determine the efficiency of heat exchangers.

- effectiveness_ntu estimates the effectiveness of a heat exchanger, crucial for performance prediction.

- `heat_transfer_coefficient` calculates the convective heat transfer coefficient, aiding in the determination of heat exchange efficiency.

- `temperature_profile` predicts temperature profiles along the heat exchanger, facilitating design optimization.

The example block illustrates the practical application of these functions using example numerical values.

Multiple Choice Questions

1. Which variable represents the specific heat capacity at constant pressure in heat exchanger calculations?

    (a) U
    (b) h
    (c) c_p
    (d) k

2. In the LMTD method, what is ΔT_{lm} used to approximate?

    (a) The overall heat transfer coefficient
    (b) The average temperature difference across the heat exchanger
    (c) The specific heat capacity of the fluid
    (d) The mass flow rate of the stream

3. What does the NTU in the effectiveness-NTU method stand for?

    (a) Nested Thermal Units
    (b) Number of Transfer Units
    (c) Non-Turbulent Utility
    (d) Nominal Thermal Uplift

4. Which number is represented by 0.023 in the Dittus-Boelter equation for turbulent flow inside tubes?

    (a) Reynolds number coefficient
    (b) Empirical constant

(c) Prandtl number exponent

(d) Nusselt number correction

5. What kind of flow configuration maximizes the temperature difference across a heat exchanger?

   (a) Parallel flow

   (b) Counterflow

   (c) Crossflow

   (d) Turbulent flow

6. How are heat transfer coefficients typically evaluated in heat exchanger analysis?

   (a) Using the LMTD method

   (b) Through numerical integration

   (c) Using empirical correlations and fluid properties

   (d) Via computational fluid dynamics

7. What role does CFD play in heat exchanger design?

   (a) It estimates the effectiveness using the NTU method.

   (b) It provides a graphical representation of flow patterns.

   (c) It offers detailed insights into temperature and velocity fields.

   (d) It provides the empirical constants for turbulent flow equations.

Answers:

1. **C:** c_p This option correctly identifies c_p as the specific heat capacity at constant pressure, which is essential in calculating heat transfer in heat exchangers.

2. **B: The average temperature difference across the heat exchanger** The LMTD (ΔT_{lm}) approximates the average temperature difference between the hot and cold fluids, facilitating energy balance calculations.

3. **B: Number of Transfer Units** NTU stands for Number of Transfer Units, a key parameter in determining the effectiveness of a heat exchanger.

4. **B: Empirical constant** The number 0.023 in the Dittus-Boelter equation is an empirical constant used in calculating the Nusselt number for turbulent flow inside tubes.

5. **B: Counterflow** Counterflow configurations maximize the temperature difference across a heat exchanger, thereby enhancing heat transfer efficiency.

6. **C: Using empirical correlations and fluid properties** Heat transfer coefficients are generally evaluated using empirical correlations that account for fluid properties and flow conditions, such as Reynolds, Prandtl, and Nusselt numbers.

7. **C: It offers detailed insights into temperature and velocity fields.** CFD provides detailed simulations of local flow and temperature patterns, considering complex exchanger geometries and material properties.

Chapter 31

Viscoelastic Fluid Flow

Introduction to Viscoelasticity

Viscoelasticity is a complex property of fluids that exhibit both viscous and elastic characteristics when undergoing deformation. The behavior of viscoelastic fluids is illustrated by their ability to dissipate energy while simultaneously storing energy, which results in distinctive flow traits in response to applied stress. Classical mechanical engineering principles and advanced mathematical modeling are employed to comprehend these unique attributes.

Governing Equations of Viscoelastic Fluids

The mathematical depiction of viscoelastic fluid dynamics necessitates a synthesis of constitutive equations that integrate both Newtonian viscous flow characteristics and Hookean elastic behavior. Such models are typically manifested in terms of tensorial equations that account for stress, strain, and time-dependent behavior.

1 Constitutive Models

Numerous models are predicated upon the differential and integral frameworks that describe stress-strain relationships. One of the prevalent models is the Oldroyd-B model, which effectively captures the linear viscoelastic response:

$$\tau + \lambda_1 \frac{\mathrm{D}\tau}{\mathrm{D}t} = \eta \left(\dot{\gamma} + \lambda_2 \frac{\mathrm{D}\dot{\gamma}}{\mathrm{D}t} \right)$$

where τ is the stress tensor, λ_1 and λ_2 represent relaxation and retardation times, respectively, η denotes the dynamic viscosity, and $\dot{\gamma}$ is the rate of shear strain. The Oldroyd-B model is instrumental in elucidating various flow phenomena inherent to viscoelastic fluids.

2 Stress Tensor and Rate of Deformation

The stress tensor σ, encompassing both the Newtonian viscous component and an elastic portion, is integral for resolving viscoelastic flow problems:

$$\sigma = -p\mathbf{I} + 2\eta \mathbf{D} + \mathbf{T}$$

where p is the pressure, $\mathbf{I}$ is the identity matrix, $\mathbf{D}$ represents the rate of deformation tensor, and $\mathbf{T}$ encapsulates the polymeric stress contribution.

Behavior under Deformation and Varying Strain Rates

The response of viscoelastic fluids to deformation is markedly influenced by varying strain rates, elucidated through constitutive adaptations such as the Generalized Maxwell Model and Power Law models.

1 Generalized Maxwell Model

The Generalized Maxwell Model characterizes the fluid's relaxation spectrum, effectively revealing time-dependent behavior through a composite structure of spring-dashpot elements. The model's viscosity function $\eta(t)$ can be expressed as:

$$\eta(t) = \sum_{i=1}^{N} \eta_i e^{-t/\lambda_i}$$

where η_i are the viscosities of discrete Maxwell elements, and λ_i are the corresponding relaxation times, with N denoting the number of elements.

2 Power Law Model for Non-Newtonian Behavior

For certain viscoelastic fluids, especially those showing shear-thinning or thickening behavior, the Power Law Model provides a robust framework:

$$\eta_{\text{eff}} = K \left|\dot{\gamma}\right|^{n-1}$$

where η_{eff} is the effective viscosity, K is the consistency coefficient, n the flow index, and $\dot{\gamma}$ is the shear rate. Here, $n < 1$ signifies shear-thinning, while $n > 1$ indicates shear-thickening behavior.

Numerical Simulation Techniques

Modern computational methods enable precise modeling and evaluation of viscoelastic fluid flows, accommodating varied flow regimes and boundary conditions.

1 Finite Element Analysis

Finite Element Analysis (FEA) remains pivotal for simulating viscoelastic fluid dynamics, offering spatial discretization capabilities necessary to resolve complex geometries and non-linear behaviors. The discretized Navier-Stokes equations incorporating the chosen constitutive model provide the computational framework.

$$\int_\Omega \frac{\partial \mathbf{u}}{\partial t} d\Omega = \int_\Omega \mathbf{f} \cdot \delta \mathbf{u} d\Omega + \int_\Omega \sigma \cdot \nabla \delta \mathbf{u} d\Omega$$

where $\mathbf{u}$ denotes velocity, $\mathbf{f}$ represents body forces, and Ω is the domain of interest.

2 CFD and Viscoelasticity

Computational Fluid Dynamics (CFD) is employed to simulate intricate viscoelastic behaviors, incorporating turbulence models and addressing numerical instabilities inherent to high-Weissenberg number flows. Modifications accommodate polymeric constitutive equations in standard CFD solvers, such as those derived from the Lattice Boltzmann Method (LBM).

$$f_i(\mathbf{x} + \mathbf{c}_i \Delta t, t + \Delta t) = f_i(\mathbf{x}, t) + \Omega_i$$

where f_i represents the distribution functions, $\mathbf{c}_i$ are the discrete velocities, and Ω_i indicates the collision operator.

Applications in Engineering Systems

Viscoelastic fluid flows are integral to several engineering systems, necessitating precise control and analysis. Typical applications include polymer extrusion, enhanced oil recovery, and biomedical devices, where understanding viscoelastic dynamics contributes to performance enhancements and operational efficiencies.

Python Code Snippet

Below is a Python code snippet that encompasses the core computational elements for viscoelastic fluid flow analysis, including constitutive model implementation, stress tensor evaluation, and numerical simulation techniques.

```
import numpy as np

def oldroyd_b_model(stress, shear_rate, lambda1, lambda2, eta,
     delta_t):
    '''
    Calculate the updated stress tensor using Oldroyd-B constitutive
     model.
    :param stress: Current stress tensor.
    :param shear_rate: Current shear rate tensor.
    :param lambda1: Relaxation time.
    :param lambda2: Retardation time.
    :param eta: Dynamic viscosity.
    :param delta_t: Time step.
    :return: Updated stress tensor.
    '''
    dstress_dt = (eta * (shear_rate + lambda2 *
     np.gradient(shear_rate, delta_t)) -
            stress) / lambda1
    return stress + dstress_dt * delta_t

def general_maxwell_model(viscosities, relaxation_times, t):
    '''
    Generalized Maxwell Model to calculate viscosity over time.
    :param viscosities: List of element viscosities.
    :param relaxation_times: List of relaxation times.
    :param t: Time.
    :return: Viscosity at given time.
    '''
```

```python
    return sum(eta * np.exp(-t / lambda_i) for eta, lambda_i in
        zip(viscosities, relaxation_times))

def stress_tensor(pressure, dynamic_viscosity, deformation_rate,
    polymeric_stress):
    '''
    Calculate the stress tensor for a viscoelastic fluid.
    :param pressure: Pressure in the fluid.
    :param dynamic_viscosity: Dynamic viscosity.
    :param deformation_rate: Rate of deformation tensor.
    :param polymeric_stress: Polymeric stress tensor.
    :return: Total stress tensor.
    '''
    identity_matrix = np.eye(deformation_rate.shape[0])
    return -pressure * identity_matrix + 2 * dynamic_viscosity *
        deformation_rate + polymeric_stress

def effective_viscosity(shear_rate, K, n):
    '''
    Calculate effective viscosity using Power Law Model for
        Non-Newtonian behaviors.
    :param shear_rate: Shear rate.
    :param K: Consistency coefficient.
    :param n: Flow index.
    :return: Effective viscosity.
    '''
    return K * abs(shear_rate) ** (n - 1)

def simulate_viscoelastic_flow(velocity_field, domain, time_steps,
    model_parameters):
    '''
    Simulate the viscoelastic flow using a numerical method (e.g.,
        FEA or CFD).
    :param velocity_field: Initial velocity field.
    :param domain: Spatial domain.
    :param time_steps: Number of time steps for simulation.
    :param model_parameters: Parameters for the constitutive model.
    :return: Updated velocity field.
    '''
    # Placeholder for simulation process, typically using FEA/CFD
    #   libraries
    for t in range(time_steps):
        # Update velocity field based on previously defined
        #   equations
        # Here using direct integration as a placeholder
        stress = oldroyd_b_model(model_parameters['stress'],
            model_parameters['shear_rate'],
                                 model_parameters['lambda1'],
                                     model_parameters['lambda2'],
                                 model_parameters['eta'],
                                     model_parameters['delta_t'])
        # Additional computational steps can be added based on
        #   specific finite element or CFD framework
```

```
    return velocity_field

# Example usage with dummy parameters
stress = np.array([[0.0, 0.0], [0.0, 0.0]])
shear_rate = np.array([[0.1, 0.0], [0.0, -0.1]])
model_params = {'lambda1': 1.0, 'lambda2': 0.5, 'eta': 1.0,
    'delta_t': 0.01, 'stress': stress, 'shear_rate': shear_rate}

updated_stress = oldroyd_b_model(stress, shear_rate,
    model_params['lambda1'], model_params['lambda2'],
    model_params['eta'], model_params['delta_t'])
viscosity_time = general_maxwell_model([1.0, 0.5], [0.2, 0.3], 0.5)
stress_tensor_val = stress_tensor(1.0, 0.9, shear_rate,
    updated_stress)
effective_visc = effective_viscosity(0.2, 1.0, 0.8)

print("Updated Stress Tensor:", updated_stress)
print("Viscosity Over Time:", viscosity_time)
print("Stress Tensor:", stress_tensor_val)
print("Effective Viscosity:", effective_visc)
```

This code defines several key functions necessary for the analysis of viscoelastic fluid dynamics:

- `oldroyd_b_model` updates the stress tensor using the Oldroyd-B constitutive model, essential for capturing viscoelastic behaviors.

- `general_maxwell_model` computes the time-dependent viscosity using a Generalized Maxwell Model to illustrate relaxation spectrum.

- `stress_tensor` evaluates the complete stress tensor, considering pressure and polymeric stress contributions.

- `effective_viscosity` calculates the effective viscosity for non-Newtonian flow using the Power Law Model.

- `simulate_viscoelastic_flow` demonstrates a placeholder for simulating flow using numerical methodologies like FEA or CFD.

The final block of code provides examples of computing elements of viscoelastic fluid flow using dummy data.

Multiple Choice Questions

1. Which of the following best characterizes viscoelastic fluids?

(a) Fluids that exhibit purely viscous behavior

(b) Solids that exhibit purely elastic behavior

(c) Fluids that exhibit both viscous and elastic characteristics

(d) Solids with high tensile strength

2. The Oldroyd-B model is primarily used to describe:

    (a) Inviscid fluids with no resistance to shear

    (b) Viscous fluids at low Reynolds numbers

    (c) Linear viscoelastic response of complex fluids

    (d) Perfectly elastic materials under deformation

3. Which term in the Oldroyd-B model equation governs the fluid's ability to return to its original shape?

    (a) $\lambda_1 \frac{D\tau}{Dt}$

    (b) $\eta \dot{\gamma}$

    (c) $-p\mathbf{I}$

    (d) $\mathbf{T}$

4. In the context of viscoelastic fluids, the Generalized Maxwell Model is used to:

    (a) Analyze fluids that follow Newton's law of viscosity

    (b) Describe the stress-strain relationship in elastic solids

    (c) Characterize time-dependent behaviors through spring-dashpot mechanisms

    (d) Model compressible gas dynamics

5. What does the flow index n signify in the Power Law Model?

    (a) The density of the fluid

    (b) The temperature dependence of fluid viscosity

    (c) The nature of the fluid's flow; shear-thinning or shear-thickening

    (d) The magnetic properties of the fluid

6. For which numerical simulation technique is viscoelastic fluid modeling particularly significant?

(a) Lattice Boltzmann Method (LBM)

(b) Molecular dynamics

(c) Mesh-free particle methods

(d) Computational Fluid Dynamics (CFD)

7. Which of the following applications would likely involve the use of viscoelastic fluid flow analysis?

(a) Extraction of metals in a smelting process

(b) Fluid flow in rocket nozzles

(c) Polymer extrusion processes

(d) Cooling water management in power plants

Answers:
1. C: Fluids that exhibit both viscous and elastic characteristics
Viscoelastic fluids have properties of both viscosity (fluid resistance to flow) and elasticity (tendency to return to their original shape).
2. C: Linear viscoelastic response of complex fluids
The Oldroyd-B model captures the linear stress-strain behavior in viscoelastic fluids, accounting for both fluid viscosity and elasticity.
3. A: $\lambda_1 \frac{D\tau}{Dt}$
This term represents the relaxation time, which is a key factor in how long after deformation a fluid takes to return to its original shape.
4. C: Characterize time-dependent behaviors through spring-dashpot mechanisms
The Generalized Maxwell Model uses spring-dashpot systems to model the time-dependent stress relaxation properties of viscoelastic materials.
5. C: The nature of the fluid's flow; shear-thinning or shear-thickening
The flow index n in the Power Law Model indicates whether a fluid exhibits shear-thinning ($n < 1$), shear-thickening ($n > 1$), or Newtonian behavior ($n = 1$).
6. D: Computational Fluid Dynamics (CFD)
CFD is widely used to simulate the complex behaviors of viscoelastic fluids, especially in response to dynamic boundary conditions and geometries.

7. C: Polymer extrusion processes
Viscoelastic flow analysis is crucial for understanding and optimizing polymer extrusion, as it involves materials that display both viscous and elastic properties.

Chapter 32

Two-Phase Flow Instability

Introduction to Two-Phase Flow

Two-phase flow refers to the simultaneous flow of two distinct phases, typically involving liquid and gas, within a conduit. This regime is prevalent in various mechanical systems and industries, ranging from chemical processing to power generation. Understanding the stability of such flows is critical, as instabilities can lead to severe operational challenges and system failures.

Stability Criteria in Two-Phase Flow

Analyzing the stability of two-phase flow involves determining the conditions under which disturbances in the flow field either decay or amplify. Key stability criteria are derived from perturbation analysis and linear stability theory.

The linear stability analysis begins with the Navier-Stokes equations, which are linearized around a basic steady-state flow solution. The perturbation equations can be represented in matrix form as:

$$\frac{\partial}{\partial t}\mathbf{q}' + \mathbf{A}\frac{\partial}{\partial x}\mathbf{q}' = \mathbf{B}\mathbf{q}'$$

where $\mathbf{q}'$ denotes the perturbation vector, $\mathbf{A}$ represents the convective matrix, and $\mathbf{B}$ corresponds to the perturbation source matrix. The eigenvalues of this equation system indicate the stability; negative eigenvalues signify stability, while positive values indicate instability.

Pattern Formation and Flow Regimes

In two-phase flow, certain flow patterns and regimes are more susceptible to instabilities. Common flow regimes such as bubbly, slug, churn, and annular flow each exhibit unique instability characteristics, typically modeled by dimensionless numbers like the Weber number (We) and the Froude number (Fr).

1 Weber Number

The Weber number quantifies the relative importance of inertial forces to surface tension forces and is defined as:

$$\text{We} = \frac{\rho v^2 D}{\sigma}$$

where ρ is the fluid density, v is the velocity, D the characteristic length, and σ the surface tension. High Weber numbers are generally associated with instabilities in annular and wispy-annular flows.

2 Froude Number

The Froude number represents the ratio of a fluid's inertial forces to gravitational forces, expressed as:

$$\text{Fr} = \frac{v}{\sqrt{gL}}$$

where g is the acceleration due to gravity and L is the characteristic length. In two-phase flow, a low Froude number suggests a dominance of gravitational effects, potentially leading to slug flow instability.

Mathematical Models of Instability

To model the instability phenomena within two-phase flow systems, several sophisticated mathematical paradigms are utilized.

1 Homogeneous Equilibrium Model

The Homogeneous Equilibrium Model (HEM) assumes that both phases are uniformly mixed and reach instantaneous equilibrium. The governing equations, under this assumption, reduce to:

$$\frac{\partial}{\partial t}(\alpha\rho) + \nabla \cdot (\alpha\rho\mathbf{v}) = 0$$

where α is the void fraction of the gas phase, ρ represents the average density, and $\mathbf{v}$ denotes the velocity vector. This model is simplistic and often used for initial stability analyses due to its reduced complexity.

2 Drift-Flux Model

The Drift-Flux Model offers more detail by considering the relative velocity between the phases. The model's momentum equation is given by:

$$\frac{\partial}{\partial t}(\rho\mathbf{v}) + \nabla \cdot (\rho\mathbf{v}\mathbf{v} + P\mathbf{I}) = \rho\mathbf{g} - \sum \mathbf{F}_{\text{interphase}}$$

where P denotes pressure, $\mathbf{I}$ represents the identity matrix, $\mathbf{g}$ is the gravitational force per unit mass, and $\sum \mathbf{F}_{\text{interphase}}$ accounts for the interphase forces. The drift-flux model can predict phase distribution more accurately than the HEM in unstable flows.

Numerical Simulation of Two-Phase Flow Instability

Accurate simulation of two-phase flow instabilities requires robust numerical techniques due to their inherent complexity. Computational Fluid Dynamics (CFD) is commonly employed, enabling validation of stability criteria and further exploration of complex flow patterns and predicted instabilities.

1 Volume of Fluid Method

The Volume of Fluid (VOF) method is widely used for simulating immiscible phases. It tracks the interface by solving the advection equation:

$$\frac{\partial}{\partial t}(\alpha) + \nabla \cdot (\alpha \mathbf{v}) = 0$$

This method is particularly effective in capturing the interface dynamics of two-phase systems, including bubble coalescence and breakup, which are critical to understanding flow instabilities.

2 Level-Set Method

The Level-Set Method is a complementary approach that represents the interface implicitly with a continuous function ϕ, solving:

$$\frac{\partial \phi}{\partial t} + \mathbf{v} \cdot \nabla \phi = 0$$

This method achieves high accuracy in tracking the interface topology across complex two-phase flow domains and supports the simulation and analysis of instability phenomena in two-phase piping systems.

Python Code Snippet

Below is a Python code snippet that encompasses the core computational elements of two-phase flow instability analysis including the perturbation analysis, calculation of dimensionless numbers, and simulation methods.

```
import numpy as np

def perturbation_analysis(A, B):
    '''
    Perform a basic eigenvalue analysis to assess flow stability.
    :param A: Convective matrix.
    :param B: Perturbation source matrix.
    :return: Eigenvalues of the system.
    '''
    # Solve the eigenvalue problem
    eigvals, _ = np.linalg.eig(B - A)
    return eigvals

def weber_number(rho, v, D, sigma):
    '''
    Calculate the Weber number.
    :param rho: Fluid density.
    :param v: Velocity of the flow.
    :param D: Characteristic length.
```

```
    :param sigma: Surface tension.
    :return: Weber number.
    '''
    return (rho * v**2 * D) / sigma

def froude_number(v, g, L):
    '''
    Calculate the Froude number.
    :param v: Velocity of the flow.
    :param g: Acceleration due to gravity.
    :param L: Characteristic length.
    :return: Froude number.
    '''
    return v / np.sqrt(g * L)

def homogeneous_equilibrium_model(alpha, rho, v):
    '''
    Simplified continuity equation for the HEM model in two-phase
    ↪  flow.
    :param alpha: Void fraction of the gas phase.
    :param rho: Average density.
    :param v: Velocity vector.
    :return: Rate of change of mass in the system.
    '''
    return np.gradient(alpha * rho) + np.gradient(alpha * rho * v)

def drift_flux_model(rho, v, P, g, F_interphase):
    '''
    Momentum balance for the Drift-Flux Model.
    :param rho: Density.
    :param v: Velocity vector.
    :param P: Pressure term.
    :param g: Gravitational force vector.
    :param F_interphase: Interphase forces.
    :return: Momentum balance equation.
    '''
    return np.gradient(rho * v) + np.gradient(rho * np.outer(v, v) +
    ↪    P) - rho * g - sum(F_interphase)

def volume_of_fluid_method(alpha, v):
    '''
    Simulation of the Volume of Fluid Method.
    :param alpha: Volume fraction of phases.
    :param v: Velocity field.
    :return: Volume fraction equation.
    '''
    return np.gradient(alpha) + np.gradient(alpha * v)

def level_set_method(phi, v):
    '''
    Simulation of the Level-Set Method for interface tracking.
    :param phi: Level set function.
    :param v: Velocity field.
```

```
    :return: Evolution of the level-set function.
    '''
    return np.gradient(phi) + v * np.gradient(phi)

# Example parameters for simulation
A = np.array([[0, 1], [-2, -3]])
B = np.array([[0, 0], [0, 1]])
rho = 1000
v = 1.5
D = 0.05
sigma = 0.072
g = 9.81
L = 0.5
alpha = 0.2
P = 101325
F_interphase = [0.01, 0.02]

# Executing functions with example values
eigvals = perturbation_analysis(A, B)
we_num = weber_number(rho, v, D, sigma)
fr_num = froude_number(v, g, L)
hem_result = homogeneous_equilibrium_model(alpha, rho, v)
drift_flux_result = drift_flux_model(rho, v, P, g, F_interphase)

print("Eigenvalues:", eigvals)
print("Weber Number:", we_num)
print("Froude Number:", fr_num)
print("HEM Result:", hem_result)
print("Drift Flux Result:", drift_flux_result)
```

This code defines several key functions necessary for the analysis of two-phase flow instabilities:

- `perturbation_analysis` performs eigenvalue analysis to determine flow stability by analyzing the perturbation matrices.

- `weber_number` and `froude_number` calculate important dimensionless numbers to characterize flow regimes and potential instabilities.

- `homogeneous_equilibrium_model` and `drift_flux_model` provide different approaches to model instabilities by simplifying or accurately depicting phase interactions.

- `volume_of_fluid_method` and `level_set_method` simulate the interface dynamics crucial for understanding two-phase flow instabilities.

The final block of code demonstrates the execution of these methods using sample parameter values, showcasing their application in a two-phase flow context.

Multiple Choice Questions

1. In the context of two-phase flow instability, what is the significance of a negative eigenvalue in the perturbation analysis?

    (a) It indicates a stable flow condition.

    (b) It represents an unstable flow condition.

    (c) It suggests a neutral stability condition.

    (d) It implies catastrophic flow failure.

2. The Weber number (We) is critical in understanding flow instability in which of the following flow regimes?

    (a) Bubbly flow

    (b) Annular flow

    (c) Slug flow

    (d) Churn flow

3. Which mathematical model assumes uniform mixing and instantaneous equilibrium in two-phase flow analysis?

    (a) Drift-Flux Model

    (b) Homogeneous Equilibrium Model (HEM)

    (c) Volume of Fluid Method

    (d) Level-Set Method

4. What role does the Froude number (Fr) play in two-phase flow systems?

    (a) It indicates thermal stability.

    (b) It quantifies wave adhesion.

    (c) It measures the influence of gravitational forces on inertial forces.

    (d) It determines chemical reaction rates.

5. Which numerical method is preferred for tracking the interface dynamics in immiscible two-phase systems?

(a) Analytical Solution

(b) Finite Difference Method

(c) Volume of Fluid (VOF) Method

(d) Euler Method

6. In two-phase flow instability, why is the Drift-Flux Model considered a more detailed approach than the Homogeneous Equilibrium Model?

   (a) It simplifies computations by assuming steady flow.

   (b) It introduces relative velocity between phases and provides detailed phase distribution.

   (c) It totally ignores interphase interactions.

   (d) It solves the problem in a higher temperature approximation.

7. Which of the following best describes the application of the Level-Set Method in two-phase flow systems?

   (a) It uses a tracer particle to track flow discontinuities.

   (b) It provides precise velocity field estimation.

   (c) It implicitly captures the interface with a continuous function.

   (d) It assumes a single-phase flow for simplification.

Answers:

1. **A: It indicates a stable flow condition** Negative eigenvalues in perturbation analyses suggest that disturbances to the flow decay over time, indicating stability within the flow system.

2. **B: Annular flow** The Weber number is particularly significant in annular flows, where the balance between inertial and surface tension forces dictates the stability of the liquid film.

3. **B: Homogeneous Equilibrium Model (HEM)** The HEM model assumes both phases are well-mixed and reach equilibrium instantaneously, simplifying the complex interactions between the phases.

4. **C: It measures the influence of gravitational forces on inertial forces** The Froude number evaluates the relative significance of gravitational forces compared to inertial forces, which is crucial for determining flow regime transitions.

5. **C: Volume of Fluid (VOF) Method** The VOF method is proficient at capturing the interface dynamics in two-phase systems, especially for complex interface phenomena like bubble coalescence and breakup.

6. **B: It introduces relative velocity between phases and provides detailed phase distribution** The Drift-Flux Model accounts for the relative motion between fluid phases, offering a more nuanced analysis compared to the simplistic assumptions made in the HEM.

7. **C: It implicitly captures the interface with a continuous function** The Level-Set Method is known for accurately representing fluid interfaces using a continuous function, aiding in detailed analysis of phase interactions within two-phase flow domains.

Chapter 33

Piping Systems Surge Analysis

Introduction to Surge Phenomena

Pressure surges, often referred to as hydraulic transients or water hammer, are critical phenomena within piping systems. These pressure oscillations occur due to rapid changes in fluid velocity, induced by pump shutdowns, valve closures, or other disruptions. An advanced understanding of these dynamics is essential to anticipate and mitigate potential adverse effects on system integrity and operation.

Governing Equations

The mathematical framework for analyzing pressure surges relies extensively on the conservation principles reflected in the continuity and momentum equations.

1 Continuity Equation

The continuity equation for a fluid element within a pipeline is expressed as:

$$\frac{\partial}{\partial t}(\rho A) + \frac{\partial}{\partial x}(\rho A v) = 0$$

where ρ is the fluid density, A the cross-sectional area of the pipe, v the fluid velocity, and x the spatial coordinate along the pipe.

2 Momentum Equation

The momentum equation in the context of an incompressible flow and one-dimensional analysis can be represented as:

$$\frac{\partial v}{\partial t} + v\frac{\partial v}{\partial x} + \frac{1}{\rho}\frac{\partial P}{\partial x} + f = 0$$

where P is the pressure, and f represents the frictional resistance term, often modeled using Darcy-Weisbach friction factor λ.

Wave Propagation in Pipelines

The propagation of pressure waves through a fluid medium is a fundamental aspect of surge analysis. The wave speed c is dictated by the fluid compressibility and pipe wall elasticity, given by:

$$c = \sqrt{\frac{K/\rho}{1 + \frac{KD}{Ee}}}$$

in which K is the fluid bulk modulus, E the modulus of elasticity of the pipe material, D the pipe diameter, and e the wall thickness.

Methods for Surge Mitigation

Mitigating hydraulic surges involves strategies to modulate pressure wave influences and reduce potential system damage.

1 Pressure Relief Valves

Pressure relief valves are employed to dissipate excess pressure, activated when thresholds exceed pre-established conditions. The dynamic response of these valves is key for effective surge control, represented by equations governing valve orifice flow:

$$Q = C_d A_v \sqrt{\frac{2(P_i - P_o)}{\rho}}$$

where Q is the discharge flow rate, C_d the discharge coefficient, A_v the valve area, P_i and P_o the inlet and outlet pressures, respectively.

2 Air Chambers and Surge Tanks

Air chambers and surge tanks provide volumetric capacity for transient pressure adjustments. The operational dynamics involve gas laws where, for ideal gases, pressure and volume relationships are defined as:

$$PV^\gamma = \text{constant}$$

In this expression, P is the pressure, V the volume of the gas, and γ the specific heat ratio.

Numerical Approaches

Analytical solutions for surge conditions in complex piping networks may become intractable, necessitating numerical methodologies such as the Method of Characteristics (MOC).

1 Method of Characteristics

The MOC transforms partial differential equations into ordinary differential equations along characteristic lines. For the standard surge equations in a pipeline, the characteristic equations are:

$$\mathrm{d}v + \frac{1}{\rho}\mathrm{d}P + f\mathrm{d}x = 0$$

$$\mathrm{d}v - \frac{1}{\rho}\mathrm{d}P + f\mathrm{d}x = 0$$

where the terms are balanced along the positive and negative characteristic directions, facilitating efficient numerical integration.

2 Finite Element Method

The finite element method (FEM) resolves surge dynamics by discretizing the pipeline into elements, providing a framework for accommodating geometric and material complexities. The system of equations is represented by:

$$Ku = f$$

where **K** is the system stiffness matrix, **u** the nodal displacement vector, and **f** the force vector, inclusive of external and internal pressure forces.

Advanced Simulation Techniques

High-fidelity simulations are indispensable for evaluating surge impacts and refining mitigation strategies. These simulations integrate computational fluid dynamics (CFD) with surge analysis to offer enhanced resolution of transient behaviors, accounting for factors like fluid-structure interaction and complex system boundary conditions.

Python Code Snippet

Below is a Python code snippet that encompasses the core computational elements for analyzing pressure surges within piping systems, derived from the governing equations of continuity, momentum, wave speed, and surge mitigation strategies.

```python
import numpy as np

def continuity_equation(rho, A, v, dx, dt):
    '''
    Solve the continuity equation for a fluid element within a
    ↪ pipeline.
    :param rho: Fluid density.
    :param A: Cross-sectional area of the pipe.
    :param v: Fluid velocity.
    :param dx: Change in spatial coordinate along the pipe.
    :param dt: Change in time.
    :return: Change in density over time.
    '''
    return -(rho * A * v * dx) / dt

def momentum_equation(rho, v, P, f, dx, dt):
    '''
    Solve the momentum equation in the context of an incompressible
    ↪ flow.
    :param rho: Fluid density.
    :param v: Fluid velocity.
    :param P: Pressure.
```

```
    :param f: Frictional resistance term.
    :param dx: Change in spatial coordinate along the pipe.
    :param dt: Change in time.
    :return: Change in velocity over time.
    '''
    return - (v * (v + dx) + (1/rho) * (P + dx) + f * dx) / dt

def wave_speed(K, rho, E, D, e):
    '''
    Calculate the wave speed for surge analysis.
    :param K: Fluid bulk modulus.
    :param rho: Fluid density.
    :param E: Modulus of elasticity of the pipe material.
    :param D: Pipe diameter.
    :param e: Wall thickness.
    :return: Wave speed.
    '''
    return np.sqrt((K / rho) / (1 + (K * D) / (E * e)))

def pressure_relief_flow(C_d, A_v, P_i, P_o, rho):
    '''
    Calculate the discharge flow rate for pressure relief valves.
    :param C_d: Discharge coefficient.
    :param A_v: Valve area.
    :param P_i: Inlet pressure.
    :param P_o: Outlet pressure.
    :param rho: Fluid density.
    :return: Discharge flow rate.
    '''
    return C_d * A_v * np.sqrt(2 * (P_i - P_o) / rho)

def surge_tank_pressure(P, V, gamma):
    '''
    Calculate the pressure-volume relationship in surge tanks.
    :param P: Pressure.
    :param V: Volume of the gas.
    :param gamma: Specific heat ratio.
    :return: Pressure-volume relationship constant.
    '''
    return P * V**gamma

def method_of_characteristics(v, P, f, dx, rho):
    '''
    Implement the Method of Characteristics for surge analysis.
    :param v: Fluid velocity.
    :param P: Pressure.
    :param f: Frictional resistance term.
    :param dx: Change in spatial coordinate along the pipe.
    :param rho: Fluid density.
    :return: Characteristic equations.
    '''
    dv1 = -(1/rho) * (P + dx) + f * dx
    dv2 = (1/rho) * (P + dx) + f * dx
```

```python
    return dv1, dv2

def finite_element_method(K_matrix, u_vector, f_vector):
    '''
    Solve the finite element method for surge dynamics.
    :param K_matrix: System stiffness matrix.
    :param u_vector: Nodal displacement vector.
    :param f_vector: Force vector.
    :return: Nodal displacements.
    '''
    return np.linalg.solve(K_matrix, f_vector - np.matmul(K_matrix,
    ↪   u_vector))

def compute_surge_impact(params):
    '''
    High-level function to compute and simulate surge impacts.
    :param params: Dictionary of parameters for simulation.
    :return: Simulated surge impact metrics.
    '''
    rho = params['rho']
    A = params['A']
    v = params['v']
    dx = params['dx']
    dt = params['dt']
    P = params['P']
    f = params['f']
    K = params['K']
    E = params['E']
    D = params['D']
    e = params['e']
    C_d = params['C_d']
    A_v = params['A_v']
    P_i = params['P_i']
    P_o = params['P_o']
    V = params['V']
    gamma = params['gamma']
    K_matrix = params['K_matrix']
    u_vector = params['u_vector']
    f_vector = params['f_vector']

    # Continuity and Momentum
    continuity_change = continuity_equation(rho, A, v, dx, dt)
    momentum_change = momentum_equation(rho, v, P, f, dx, dt)

    # Wave Speed
    wave_sp = wave_speed(K, rho, E, D, e)

    # Pressure Relief
    relief_flow = pressure_relief_flow(C_d, A_v, P_i, P_o, rho)

    # Surge Tank
    surge_tank_const = surge_tank_pressure(P, V, gamma)
```

```python
# Method of Characteristics
char_equation = method_of_characteristics(v, P, f, dx, rho)

# Finite Element Method
nodal_disp = finite_element_method(K_matrix, u_vector, f_vector)

return {
    "continuity_change": continuity_change,
    "momentum_change": momentum_change,
    "wave_speed": wave_sp,
    "relief_flow": relief_flow,
    "surge_tank_constant": surge_tank_const,
    "characteristic_equation": char_equation,
    "nodal_displacement": nodal_disp
}

# Example parameters for simulation
params = {
    'rho': 1000,
    'A': 0.1,
    'v': 1,
    'dx': 0.01,
    'dt': 0.001,
    'P': 100000,
    'f': 0.02,
    'K': 2.2e9,
    'E': 200e9,
    'D': 0.5,
    'e': 0.005,
    'C_d': 0.6,
    'A_v': 0.05,
    'P_i': 110000,
    'P_o': 100000,
    'V': 1.0,
    'gamma': 1.4,
    'K_matrix': np.array([[2, -1], [-1, 2]]),
    'u_vector': np.array([[0], [0]]),
    'f_vector': np.array([[1], [0]])
}

# Simulate surge impacts
results = compute_surge_impact(params)

# Print results
for key, value in results.items():
    print(f"{key}: {value}")
```

This code defines several key functions necessary for the implementation and analysis of pressure surge phenomena in piping systems:

- `continuity_equation` function calculates the change in density for a fluid element.

- `momentum_equation` evaluates changes in fluid velocity due to pressure and friction.
- `wave_speed` computes the speed of pressure waves in the pipeline.
- `pressure_relief_flow` calculates discharge flow through pressure relief valves.
- `surge_tank_pressure` describes the gas pressure-volume relationship.
- `method_of_characteristics` implements this numerical method for surge analysis.
- `finite_element_method` uses FEM to solve surge dynamics by discretizing the pipeline.
- `compute_surge_impact` ties all computations together to simulate surge impact metrics.

The final block of code provides an example of simulating surge effects using these computational functions, with all results being printed for inspection.

Multiple Choice Questions

1. What is the primary cause of pressure surges, commonly known as water hammer, in piping systems?

    (a) Gradual changes in fluid velocity

    (b) Rapid changes in fluid velocity

    (c) Variations in fluid temperature

    (d) Gradual increase in fluid density

2. Which equation is essential for analyzing pressure changes in a pipeline due to a surge?

    (a) Bernoulli's equation

    (b) Navier-Stokes equation

    (c) Continuity equation

(d) Energy equation

3. In surge analysis, which component of the fluid-structure interaction is represented by the wave speed c?

    (a) Viscosity of the fluid

    (b) Temperature of the fluid

    (c) Compressibility of the fluid and elasticity of the pipe wall

    (d) Density of the fluid and cross-sectional area of the pipe

4. Which mitigation strategy involves using components to provide volumetric capacity for transient pressure adjustments?

    (a) Pressure relief valves

    (b) Air chambers and surge tanks

    (c) Increased pipeline diameter

    (d) Pump acceleration controls

5. What is the role of the Method of Characteristics (MOC) in the context of surge analysis?

    (a) To transform partial differential equations into linear equations

    (b) To transform differential equations into a set of algebraic equations

    (c) To transform partial differential equations into ordinary differential equations along characteristic lines

    (d) To transform algebraic equations into partial differential equations

6. When analyzing surge conditions using the finite element method, what does the matrix **K** represent?

    (a) The system dynamic matrix

    (b) The system stiffness matrix

    (c) The system weight matrix

    (d) The system pressure matrix

7. Which of the following is NOT a factor considered in high-fidelity simulations for evaluating surge impacts?

(a) Fluid viscosity variations

(b) Fluid-structure interactions

(c) Temperature fluctuations

(d) Complex system boundary conditions

Answers:

1. **B: Rapid changes in fluid velocity** Pressure surges, or water hammer, occur primarily due to rapid changes in fluid velocity, such as those occurring from pump shutdowns or sudden valve closures.

2. **C: Continuity equation** The continuity equation is vital for analyzing pressure changes in pipelines, as it accounts for the conservation of mass during transient conditions in the system.

3. **C: Compressibility of the fluid and elasticity of the pipe wall** Wave speed c is a key factor in surge analysis, influenced by the fluid's compressibility and the pipe wall's elasticity, determining how pressure waves propagate.

4. **B: Air chambers and surge tanks** These components provide the necessary volumetric capacity to accommodate transient pressure changes and mitigate surges in a pipeline.

5. **C: To transform partial differential equations into ordinary differential equations along characteristic lines** The MOC is used to simplify the complex partial differential equations governing surges into ordinary differential equations for easier solution.

6. **B: The system stiffness matrix** In the finite element method, $\mathbf{K}$ represents the system stiffness matrix, critical for depicting the relationships between nodes in terms of displacements and forces.

7. **C: Temperature fluctuations** High-fidelity surge simulations focus on factors like fluid-structure interactions and complex boundary conditions but typically do not consider temperature fluctuations as primary surge factors.

Printed in Great Britain
by Amazon